To Theo and Milo

Human Identity and Identification

Few things are as interesting to us as our own bodies and, by extension, our own identities. Recent years have brought a growing interest in the relationship between the body, environment and society.

Reflecting upon these developments, this book examines the role of the body in human identification, in the forging of identities and the ways in which it embodies our social worlds. The approach is integrative, taking a uniquely biological perspective and reflecting on current discourse in the social sciences. With particular reference to bioarchaeology and forensic science, the authors focus on the construction and categorisation of the body within scientific and popular discourse, examining its many tissues, from the outermost to the innermost, from the skin to DNA. Synthesising two traditionally disparate strands of research, this is a valuable contribution to research on human identification and the embodiment of identity.

REBECCA GOWLAND is a lecturer in Human Bioarchaeology in the Department of Archaeology at Durham University. Her current research interests lie in health and demography of the past, skeletal ageing and age as an aspect of social identity, social perceptions, care and treatment of the physically impaired in the past and the interrelationship between the physical body and social identity.

TIM THOMPSON is a reader in Biological and Forensic Anthropology at Teesside University, where he also acts as a consultant forensic anthropologist. Most of his research examines the effects of burning on the skeleton and how this can help to interpret the context of death. He is also interested in the relationship between identity and identification, modification of the body in the modern context and the role of forensic anthropology in the world at large.

Human Identity and Identification

REBECCA GOWLAND
Department of Archaeology,
Durham University

TIM THOMPSON
School of Science and Engineering,
Teesside University

CAMBRIDGE
UNIVERSITY PRESS

CAMBRIDGE UNIVERSITY PRESS
Cambridge, New York, Melbourne, Madrid, Cape Town,
Singapore, São Paulo, Delhi, Mexico City

Cambridge University Press
The Edinburgh Building, Cambridge CB2 8RU, UK

Published in the United States of America by Cambridge University Press, New York

www.cambridge.org
Information on this title: www.cambridge.org/9780521885911

First published 2013

Printed and Bound in the United Kingdom by the MPG Books Group

A catalogue record for this publication is available from the British Library

Library of Congress Cataloguing in Publication data
Gowland, Rebecca, author.
  Human identity and identification / Rebecca Gowland, Durham University,
  Tim Thompson, School of Science and Engineering, Teesside University.
    pages cm
  Includes bibliographical references and index.
  ISBN 978-0-521-88591-1 (hardback) – ISBN 978-0-521-71366-5 (paperback)
  1. Forensic anthropology. 2. Identification. 3. Human body. 4. Identity
  (Psychology)   I. Thompson, Timothy James Upton, author. II. Title.
  GN69.8.G69   2013
  301–dc23   2012033218

ISBN 978-0-521-88591-1 Hardback
ISBN 978-0-521-71366-5 Paperback

Contents

Acknowledgements

From its relatively simple conception, this book rapidly became a huge undertaking, one which ended up heading in a different direction to originally thought. We couldn't have completed this work without the help and support of our colleagues and friends. We would also like to thank past and present students for their thought-provoking and challenging discussions.

In particular, we would like to thank those who read through draft versions of chapters of this book and gave their time and valuable comments. Principally we'd like to thank Prof Tracy Shildrick and Dr Paul Crawshaw (School of Social Sciences and Law, Teesside University), Melanie Brown (School of Science and Engineering, Teesside University), Dr Susan Peake (Teesside University Business School), Dr Zoe Crossland (Columbia University) and Dr Sarah Semple (Department of Archaeology, Durham University) for their input on various chapters and Jessica Cooney (University of Cambridge) for information on prehistoric finger fluting. We would also like to thank Gaynor Western (Ossafreelance) for her copy-editing and wide-ranging and insightful comments across the book. Thanks also for the copy-editing of Alejandra Gutierrez. Naturally, the content and tone of this book is solely attributable to the authors. Tim is also grateful to the Technology Futures Institute at Teesside University which gave him time to work on this book over the past few years.

We have been able to use some great images in this book, so thanks to Dr Cathy Gaither (Metropolitan State College of Denver), Dr Anwen Caffell (Durham University) and Ian Parker (Teesside University) for giving us access to their collections.

Thanks to Carole Gowland for supplying emergency childcare, without which this book would have taken even longer to write. At Cambridge University Press we'd like to thank Katrina Halliday and Megan Waddington for their patience and support throughout the process of devising and writing this book.

1

Introduction

The body is not a beginning. It is not a starting point.

(Cream, 1994: 2)

1.1   Introduction and some historical context

Study of the physical body has a long history, with the first recorded anatomical dissections performed in Alexandria during the third century BC (Sawday, 1995; Carlino, 1999). Anatomical knowledge of the human body in the Western world was then further expanded by Galen and his followers in the second century AD (Sawday, 1995). Galen stated that students of the human body should learn it in a specific order: from the innermost to the outermost (though there is no mention of the skin) (Connor, 2004). Later, in AD 1543, Andrea Vesalius published his beautifully illustrated *De Humani Corporis Fabrica Libri Septem*, which marked a turning point in the study of human anatomy. From this point onwards we see a dramatic increase in the number and detail of visual representations of the human body in drawings and carvings (see Rifkin et al., 2006 for many examples and discussion). The human body continues to fascinate and challenge researchers and students. As a consequence of this continuous interest in ourselves, scholars have written vast amounts about the body; about its form, its function, its structure and its evolution; how it works and how it fails; how it changes and responds and how it maintains balance and homeostasis throughout the life course. Few things are as interesting to as many people as our own bodies and, directly related to this, our own identities.

This book is about the body: its role in human identification (e.g. in forensic and archaeological contexts), in the forging of identities and the multitude of ways in which it embodies our social worlds. In recent years academic discourse regarding the body has developed a more nuanced understanding of the relationship between body, environment and society. This, therefore, seems an appropriate time to reflect on these developments in relation to those disciplines engaged with human identification and the embodiment of identity. Here, we take a holistic view of the topic and discuss the materiality of the body in relation to current discourse within the social and biological sciences and its construction and categorisation within scientific and popular discourse, with particular reference to bioarchaeology and forensic science. With the exception of anthropology, the body remained largely absent from the social science literature, relegated to the status of a biological vessel in which the human social agent happens to reside (Turner, 1991; Shilling, 1993; Jackson & Scott, 2002); the body as an organic system was either allocated to biomedical disciplines, or as Turner has stated, viewed as 'an environmental constraint' (1991: 7–8). Textbooks on 'the body' generally comprise densely descriptive, heavily illustrated anatomical tomes, with little consideration given to the powerful influence of social context. The science–theory, mind–body divide meant that the 'biological' aspects of the body were the preserve of those engaged with scientific discourses (such as human identification) while social identity as a culturally specific and historical construct was a distinctive field of study in which the physical body was viewed as a largely passive 'absent presence' (Shilling, 1993).

With the pioneering work of authors such as Turner (1996) and Shilling (1993), the role of the physical body in social identity and the dialectical relationship between the two became a fresh focus of study. Flesh and blood for the first time featured in the social science research agenda. The literature in this area has blossomed since the 1990s with a number of influential books and articles exploring the active role of the physical body in human interaction and how the body is, in turn, moulded by society. A testament to this burgeoning subject was the founding of the journal *Body and Society* in 1995. While this new body-centred discourse has been highly influential within the social sciences, it has had a limited impact within the 'harder' biological disciplines, such as anatomy, biomedicine, forensic science, biometrics and genetics. It is worth noting that within anthropology the body has long been a focus of study, albeit in descriptive physiological categorisations or as a site of social mediation within ethnographic studies. Within archaeology, the embodiment of identity has received a considerable amount of attention in recent years, and some of this discourse has engaged with the body as a physical entity rather than simply as a passive clothes horse for material culture (e.g. Meskell, 1999; Joyce, 2005, 2008; Gowland & Knüsel, 2006; Sofaer, 2006; Borić & Robb, 2008; Rebay-Salisbury et al. 2010). The significance of the physical remains of the body as an essential source of data for reconstructing past lifeways has been a focus of study in bio-archaeology for many years. The more explicit theorisation of these remains as the physiological embodiment of social processes and integration with social theory has only surfaced more recently, however (e.g. Gowland & Knüsel, 2006; Sofaer, 2006; Knudson & Stojanowski, 2009).

These subjects all deal with the physical matter of the body and, whether explicitly acknowledged or not, social identity. Usually the body is treated as a universal; characteristics such as sex, age and 'race' are considered solely in biological terms. Much of this research still occurs in isolation from the social science literature concerning social identity and embodiment. Consequently, terms such as 'gender' and 'ethnicity', when they do infiltrate the scientific arena, are often poorly defined or employed incorrectly (for a discussion of this see, for example, Walker & Cook, 1998; Shim, 2005). The techniques whereby 'biological' characteristics such as sex and age are assigned to individuals within a human identification context (e.g. forensic practice, anthropology, archaeology) are described in objective, scientific terms. This cloak of objectivity, however, masks a multitude of uncertainties, not all accurately represented in the stated results. Much of this uncertainty arises because of the difficulty in mapping or describing statistically the spectrum of human variability. But this is not the whole picture. The science of human identification in all of its guises has had a long and chequered history: scientists practise their arts through the lens of their own historically situated culture and identity within it (Lewontin, 1991). Since the 1990s in particular, the conception of the science of biology as pure and objective has been increasingly questioned, particularly in relation to feminist discourse (e.g. Schiebinger, 1986; Laqueur, 1990; Spannier, 1995). For example, Spannier has argued that ultimately scientists present 'a partial vision skewed by invisible biases' (1995: 3). Science strives for objectivity and methods are tailored accordingly, but scientific endeavour has a long-established 'inherent sociality' in its construction and practice (Lambert & McDonald, 2009: 5). As Sheldon observes, 'Objective truth is a scientific aspiration but sociologically speaking it remains an impossibility' (2002). Schiebinger also expresses these sentiments when she states that 'neither science nor transhistorical bodies exist apart from culture' (2004: xiii). The body is in reality open to many different and competing interpretations (Crossland, 2009b).

In this book, we discuss the way in which the body has been conceptualised in much of the human identification literature as distinct from the concerns of social theory and the way in which our biological tissues are saturated and shaped by our social environment. The key aim of this book is to examine the different tissues of the human body and the way in which these contribute to human identification and identity research by synthesising and integrating these two traditionally disparate strands of research on the human body. Previous volumes that have attempted such an integrated approach have often focussed on one aspect or biological structure of the body (e.g. the skin or the skeleton). This book is unique in that it examines each layer from the outermost to the innermost, from the macro to the micro level. Here, we explore the tension between the social and biological disciplines in relation to the various bodily structures and examine the way in which characteristics of the body have been harnessed within human identification contexts to infer aspects of identity.

1.2   Human identification: historical context and modern applications

Now, as in the past, discussion of the biology of the body invariably leads to the application of that understanding in a more applied context, that of human identification. As already stated, in this book we focus on the physical tissues of the body as these are the raw materials those within the human identification sciences work with, though from differing academic contexts and perspectives. So we must consider the body first, as it provides a context for all subsequent discussions, and it allows us to state why this topic is of such importance and why we must discuss it fully.

When we consider human identification, perhaps the first question to pose should not be how do we do it (discussed in detail in the following chapters), but rather why do we do it? Why do we study the body in order to identify people? As Williams and Johnson note, 'There are instances of practice which use the body in ways that are not necessarily medical or surgical but that are designed to render it observable and amenable to control' (2008: 25). Throughout this book we will refer to 'human identification contexts'. We mean by this, for the most part, those situations in which the identity of an individual is assessed or described by a third party using scientific techniques. This situation may arise for several reasons, including: the individual is dead and his or her identity unknown (e.g. in archaeological and forensic contexts), the individual is living but unable to provide that information (e.g. an infant or an unconscious individual) or proof of identity is required (e.g. often using biometric technologies). Identifications of this sort occur in a wide range of disciplines, yet in all cases a successful identification is one in which the biological profile ascribed by the scientist closely matches the culturally understood identity categories. Within biomedical contexts, identities such as age, sex, ethnicity and so forth are significant for, amongst other things, epidemiological studies and the characterisation of disease processes. Within bioarchaeology, such identifications are fundamental to our understanding of the structure of past societies, as well as past human interactions and environments. Within a forensic context, human identification is necessary for establishing both victim and criminal identifications and, more recently, for resolving issues of national security.

If we wish to think about human identification as a fully fledged scientific discipline, it is worth considering its origins, some of which reside in the more dubious pseudoscientific Victorian disciplines of physiognomy and phrenology. These studies were driven by, amongst other things, an interest in criminology, whereby physical appearance and moral character were closely linked (Twine, 2002: 67). This is also referred to as 'biocriminology', and disciplines such as criminology remain preoccupied with the body (Wright & Miller, 1998; Twine, 2002; Walby & Carrier, 2010). Physiognomy still permeates a great deal of the underlying assumptions concerning identities such as 'race', class and gender:

> [T]he belief that you can read the character of another from their appearance is an historically pervasive phenomenon. Actual written treatises on physiognomy date back to at least Aristotle's (384–322 BC) and his teacher Plato's (427–348 BC) theory linking physical beauty with moral goodness. (Twine, 2002: 69)

Physiognomy formed part of the intellectual and political landscape of the Victorian era when science and pseudoscience were employed to naturalise social groupings. Work in this vein was very similar to that of well-known Italian criminal anthropologist Cesare Lombroso, who re-emphasised criminality as physiognomically written on the body, for example, the view of female criminals as having thicker-than-average jaws (Lombroso & Ferrero, 1895). Lombroso referred to the body in three key and interrelated ways: as the criminal body, the punishable body and the social body (Walby & Carrier, 2010). The first of these three is associated with 'spotting' criminality from the physical form and is the approach we most associate with him. He also argued, however, that the physicality of the body should be studied and measured as a means of assigning punishment for crimes. This was termed 'judicial anthropometry', as it assumed that the body would reveal the dangerousness of an individual and that in turn should relate to the seriousness of the punishment (Walby & Carrier, 2010). Further, as Crossland notes, the increase in photographing criminals at that time afforded a form of 'mute testimony', a visual means of recording and sharing the supposed physicality of criminality (2009b). As Twine discusses:

> [I]nferiorizations of others along lines of age, gender, class, 'race' and species are complex and different but there is a commonality in that they all draw upon a physiognomic marking of a body

that makes unsubstantiated and generalized claims upon the subjectivity of the (human) being in question. Aged skin and senility, black skin and criminality, such stereotypes still inform the social scene. We continue to give life to specific physiognomic and phrenological language when we speak of people as 'high or lowbrow', 'thick-headed' or 'thick-necked'. Moreover, links between appearance and identity are often the subtext to many contemporary issues related to self. (2002: 82)

Throughout history, pathological conditions marking the external surface of the body have been thought to reflect the inner moral or spiritual degradation of the sufferer; this is particularly apparent in early medieval accounts of disease, care and treatment (e.g. see Rawcliffe, 2006). For example, external blemishes or defects would make a man unworthy of holding high office, thus making an explicit link between physical appearance and status (Crawford, 2010: 96). Interestingly, even scientific interpretations of Anglo-Saxon skeletal remains centuries later draw on physiological characteristics to infer moral and intellectual character. A striking example of this is the skeletal report on the Anglo-Saxon cemetery of Broadchalke, Kent, in which the physical anthropologist writes of one skeleton:

Many of his features are effeminate. He reproduces characters which one can identify amongst those living round us. His head is large; the volume of his brain I estimate at 1600cc, about 120cc above the modern average … These are conditions we do not meet amongst primitive races … In this community we meet not a robust strong-limbed warrior, but a big-brained man who may well have been statesman, philosopher, poet, or clergyman. (Keith, 1925: 98)

The association between external appearance and inner character is also apparent when one examines the attitudes towards the poor in industrialised England. The wretched physical condition of the working classes in the early nineteenth century was seen as reflecting their overall moral inferiority to the higher classes rather than as stemming from their appalling working conditions, lack of education and poverty (Chapter 2). This link between physiology and intellectual and moral characteristics culminated in eugenic thought and reached its pinnacle of horror in the Nazi gas chambers. Even today the link between our outer and inner selves persists; tall people are more likely to receive promotion at work than short people; beauty (and youth) is revered and those so graced placed on a pedestal while perceived ugliness is socially debilitating. We are an ocular-centric species and, though entirely culturally specific, the interpretation of behavioural characteristics from the physical form is deeply imbued within our psyche. As we shall see throughout this book, while the technologies and language for describing human physiognomy may have changed, some of the underlying assumptions regarding biology and social behaviour have not.

1.2.1   Human identification disciplines

Many of the identification sciences serve the judicial and medico-legal context through the identification of both the living and the dead, and as Christensen and Crowder note: 'it is clear that science and the law continue to interact and interrelate' (2009: 1215). Examples discussed in this book include the relationship between DNA identification and the law (Chapter 6) and organ transplantation and the status of 'brain death' (Chapter 4). Many differing requirements exist within this broad area of the forensic sciences. For example, we might consider the identification of unknown individuals who cannot or will not identify themselves. In this context, a person may need to be identified through a tiny fragment of his or her physical being and the requirement here tends to link a person (living or dead) to a site of criminal activity or to a particular location. While within Western discourse we conceive the body as a bounded entity, this type of forensic analysis is predicated on the fact that in actuality our boundaries are permeable; our bodies imbibe and excrete, they exchange and shed, leaving an invisible corporeal trail wherever we go. As Bildhauer observes: 'Despite the usefulness of the model of the body as a separate, enclosed unit, then, this view is not at all obvious, and instead needs a lot of cultural work to be upheld' (2006: 3).

In another context, identification may involve the examination of a representation of a person, for example, an image or CCTV sequence. Examples of these various contexts are provided in a number of texts, including Thompson and Black (2007). Regardless of the specific details, Timmermans argues that for identification to be successful in this context, a connection needs to be made between the physical presence of the body and an identity (2006). He notes that a name is a sign of life, not death, and so the implication is that other means of identification are required for the deceased. The myriad identification techniques are usually separated into three categories depending on the strength of identification they provide. Primary techniques, including DNA and fingerprints, are viewed as providing absolute proof of identity. This tenet will be repeatedly tested throughout this book. Secondary techniques are those which are insufficient on their own but adequate when used in combination with other techniques, such as visual inspection and blood group. Tertiary techniques are those which can only really support other primary or secondary techniques and include personal effects and descriptions (see Thompson & Puxley, 2007, for more detail on this). Ultimately though, because 'people are entangled in multiple bureaucracies during their lives … it is difficult to die anonymously' (Timmermans, 2006: 49).

It is worth noting briefly that decomposition can severely affect the identification process. While it is easy to say that a given identification discipline focusses on the deceased, the biological decay of the body influences the nature and philosophy of the techniques used and the discipline itself. Specific details of the process of decomposition can be found elsewhere (such as in Haglund and Sorg's volume [1997]) but the process involves the cessation of basic cellular functioning, which halts the body's ability to maintain homeostasis. Thus the enzymes from within the body's cells and the bacteria in the gut can spread around the soft tissues of the body, causing colour change, swelling and ultimately disruption of the integrity of the body. Timings for this vary according to environmental conditions. This natural process is common to us all, unless certain factors (such as freezing, embalming and so on) are applied to the body to slow down this process. As will be seen in the subsequent chapters of this book, a great many facets of identification and identity depend on or originate from these soft tissues, and thus their destruction has a significant role to play in many settings.

There are, of course, other non-criminal modern contexts in which human identification is a necessity, such as when dealing with mortgages and wills, but identification here tends not to focus on the body, but rather on documentation and paperwork. In daily life, people are required to categorise and identify themselves on a regular basis for bureaucratic purposes. Sex, ethnicity and age are the most frequently requested aspects of identity ascription (see following chapters for greater exploration of this).

Like those working in the forensic sphere, anthropologists are concerned with both the living and the dead. By contrast, the anthropological literature has drawn a great deal of attention to the culturally constructed nature of bodies, as well as cross-cultural variation in identity construction. In terms of human identification, anthropologists of living societies need to make interpretations concerning age, sex and status of their subjects, even in circumstances in which these have little meaning for the society under observation. These interpretations are located within their own Western experiences of the terms and, as in a forensic context, individuals are identified as such primarily on the basis of physical appearance.

Biological anthropologists and human bioarchaeologists are also concerned with both human identification and identity, this time interpreted from the physical human remains of our ancestors. Most often the skeleton is the primary unit of analysis, with scientists using both macroscopic methods and biomolecular techniques to study it (see Chapters 5 and 6). The techniques used in the study of the ancient dead are not dissimilar to those used by forensic anthropologists for assessing the more recently deceased. The techniques themselves are produced from observing variation in physical features amongst 'known' populations, for which historical documents relating to age, sex, class and ethnicity are available, sometimes also referred to as 'identified skeletal collections'. The observed biological variation in these known populations is described statistically and an identification technique is produced and then applied to the ancient skeletal remains. This sounds straightforward, but, as many archaeologists have observed previously, it involves a whole host of assumptions concerning the universality of these physical features. Bioarchaeologists are well aware that the skeleton is very far from universal; indeed it is the skeleton's very plasticity that is harnessed to infer information about past living environments. Yet, in terms of the basic categories of identification, such concerns regarding the uniformitarian assumptions inherent in many techniques are generally brushed aside (see Chapter 5).

1.2.2   *Biometric identification*

Biometric identification is a rapidly developing field, and industry and politicians are keen to rapidly utilise this seemingly efficient and cost-effective identification solution to solve issues of national security and identity verification. Now, fingerprints are not reserved for criminals but are required to facilitate travel, or, as in the United States, in order to receive welfare. Schoolchildren provide fingerprints in schools, while iris scans, voice recognition software and so forth are all used to identify people in a variety of situations. Despite the controversy around the collection and use of biometric information from human bodies it is important to remember that 'biometrics are unique identifiers but they are not secrets' (Schneiner, 1999). Indeed, most of the key biometric features are clearly visible to all. As has just been noted, the development of this field is hotly debated. Thurtle and Mitchell express their concern at the idea that sees 'data made flesh', which they argue is a troubling trend in science in which bodies are losing out to abstract notions of information (2004: 3). Within the biometric community, the phrase 'identity management' has been adopted (Fisher, 2008), likely to reflect the fact that you are no longer passively carrying your uniqueness, but that you need to deploy and maintain it too. A failure in our identity management may lead to another person using our identity. 'Identity theft' and 'identity fraud' are becoming an increasing concern with a recent report showing that 61 per cent of Britons were concerned about identity fraud (Fisher, 2008). It is important that measures are taken to reduce such concerns and threats. Fisher states that 'stronger, robust authentication is crucial in a joined up world where information is shared' (2008: 9). In other words, the use of an increasing range of biometric identifiers from the body.

1.3   Boundaries of identity and identification

That the body and mind/soul are fundamentally distinct entities has ancient antecedents in Western medicine with traces as far back as Hippocrates and Aristotle (Scheper-Hughes & Lock, 1987). It was not until the sixteenth century that Descartes articulated the mind–body separation so distinctly: *Cogito, ergo sum*: I think therefore I am. As discussed by numerous authors, this Cartesian separation has shaped the conception and approach of Western sci-entific tradition to the body, which has been conceived of as almost entirely divorced from the mind and emotions. Within this framework the body becomes ahistorical and acultural, so that it is overwhelmingly perceived as a universal, fixed, purely physiological entity. In a series of papers, Nancy Krieger notes that in clinical medicine aspects of identity such as 'age', 'sex/gender' and 'race/ethnicity' are the holy triumvirates of epidemiological studies. Incidences of infectious, cardiovascular and other diseases are all looked at in relation to these social variables. The categories themselves are, however, rarely theo-rised or problematised and the theoretical underpinning of this choice of cat-egories is almost never developed. Socio-economic status is, of course, another prime variable in health statistics, with poor health across a number of disease categories strongly associated with lower-class groups (Krieger & Fee, 1996; Krieger, 2003, 2005; Krieger & Davy Smith, 2004).

That there is an objective truth to be uncovered, recorded and 'mapped' about the human form from the DNA, bones and external morphology is the source of considerable scientific endeavour. Indeed, some sciences of the body have taken an alarmingly deterministic turn, seeking to reduce individuals and societies to the sum of their DNA (see Lewontin, 1991) or biometric proportions. Eugenistic terms such as 'gene therapy' and 'designer babies' are infiltrating popular con-sciousness and justifications are framed in terms of 'informed choice' (Chapter 6). Furthermore, in the face of 'threats to national security', there has been an increasing drive towards biometric technologies which seek to reduce aspects of the human form to numerical sequences (Kruger et al., 2008). What could be more objective or immutable than a numerical 'code'? Within this epistem-ology the body is absolute and knowable. This has prompted Kruger et al. to argue that 'the digital extraction of information from body parts, fragments and substances facilitates a de-personalization or de-humanization effect that enables security regimes to collect, analyse and disseminate information by depoliticizing the body and denying its sociality' (2008: 116).

Feminist critiques of science initially reinforced this separation of the biological body and social identity in the 1970s. Grounded in Cartesian logic, the historical specificity of embodied experiences was emphasised (e.g. gender, ethnicity and class) while the body itself retained its status as a universal (Oudshoorn, 1994). The existence of a pre-discursive, natural body draped by culture was an important device for critiquing the biological determinism of sexism and racism (see Chapter 2). Since the 1990s, however, there has been an increasing critique of this stance (e.g. Scheper-Hughes & Lock, 1987; Turner, 1991; Shilling, 1993). A number of authors have also discussed the way in which the biologised body of Western medicine is a relatively recent and certainly not ubiquitous phenomenon (e.g. Scheper-Hughes & Lock, 1987; Krieger & Davy Smith, 2004). The marginalisation of the physical body in academic discourse on social identity has been discussed in detail by researchers who examine the way in which it has been perceived as 'an inert mass', imbued with cultural meaning, but not, of itself, active in the construction of this meaning. Budgeon argues that instead of being perceived as objects, bodies should be thought of as '*events* that are continually in the process of becoming – as multiplicities that are never just found but are made and remade' (2003: 50). Bodies both project and contain our social and biological identities, revealing aspects of our biographies and possible future trajectories. As Krieger and Davy Smith observe, emphasis on the more literal nature of embodiment has the potential to generate new insights into the way societal conditions shape the body's tissues (2004).

Here we discuss the influence of aspects of identity on our ability to successfully identify an individual; this leads us to an exploration of human identity, what this might be and how identity and identification mix and coalesce. Williams and Johnson, in their critique of the use of technologies in modern policing, argue that the notion of the 'individual as the basic unit of analysis and scrutiny is a preoccupation … of the operational practice of criminal investigations' (Williams & Johnson, 2008: 20). Furthermore, 'the growth of modern policing is inextricably linked to the development and deployment of methods of human individuation and classification' (Williams & Johnson, 2008: 19). Beyond the examination of the individual, the emphasis shifts to putting individuals into classes and categories necessary in attempts to co-ordinate methods of social control (Williams & Johnson, 2008). While there is overlap among the disciplines which employ methods of human identification, the reason for the identification is significant because it governs the specific technique that can and will be applied.

1.4   Structure of the book

We discuss the various tissues and structures of the body as described in the anatomical and human identification literature and integrate this with social discourse concerning the physicality of the body in the social sciences. One of the aims of this book is to synthesise and bridge these mostly disparate discourses on the human body in order to move the human identification sciences towards a more holistic and integrated approach. The book is structured according to the layers of the body; working from the outside inwards, from the skin to the bones, we analyse and peel back the various tissues down to the biomolecular level, to what has been described as 'the blueprint for life' itself, DNA. We discuss the techniques of identification as applied to these different bodily structures and the way in which the body becomes constructed and enmeshed within aspects of our social worlds, those of our ancestors and within our scientific traditions. Each chapter addresses a number of different themes within the social identity literature and their impact on the particular organ or tissue discussed, namely: gender, ethnicity, age and class. These ways of categorising individuals both biologically and socially have been critiqued and discussed widely within the academic literature and some of these debates are visited here in relation to the body's tissues.

To avoid repetition throughout the book, Chapter 2 provides a brief introduction to the way in which each of these categories of human identity have been conceptualised in the social and biological literature. We provide a brief history of bodily categorisations, the history of their role in the process of human identification and the problematic nature of these constructs. We discuss the concepts of human identification and identity and look at the ways in which they are linked together in such an intimate way and how the study of one aspect of the two reveals much about the other.

Our third chapter examines the integumentary system, or the skin. As the most external and visible aspect of our body, the skin has always held a central role in human identity and identification. We explore the concept of friction ridge identification (e.g. fingerprints) in relation to the underlying biology of their formation and its lasting pre-eminence in terms of biometric identification. We also examine the significance of the skin in terms of our own perceived identities, especially in relation to age, gender, ancestry and pathologies; we observe the interaction of biological and social worlds at the surface of our bodies in the orchestration of identity formation and maintenance. Our fourth chapter encompasses the deeper soft tissues. Thus we examine the importance of key soft tissues in human identification (such as blood, fat, the vascular system and organs, including the eyes) and again seek to explore the interactions between the biological and social worlds in respect to these structures and substances. Here we touch upon the boundedness of embodied identity and how this Western viewpoint can be challenged through the exchange of fluids and organs from one person to another and the impact of this for the identities of the donor and the recipient. We also examine the increasing significance of body fat, the scourge of the Western world and a socially mediated bodily substance. Our fifth chapter examines the human skeleton and deconstructs both its biology and sociology for the purpose of answering our initial line of enquiry, that is, what your body reveals about you. The human skeleton is particularly significant in many forensic and bioarchaeological contexts because of its resistance to decay relative to the soft tissues. It has therefore been the subject of considerable academic scrutiny in relation to its ability to reflect social identity and to aid in human identification contexts. We discuss the historical importance of the human skeleton in the categorisation of individuals from craniometric and morphological techniques from the seventeenth century to the present.

Our sixth chapter zooms down to the microscopic scale to examine the implication of the biomolecular view of our bodies for identification and identity. This includes our DNA but also the chemical components of our tissues in terms of isotopic ratios. This chapter is particularly pertinent since considerable media and scientific focus falls on these, the very smallest components of our bodies. But even here, wrapped as they are in the hard sciences, we see the significance of a social perspective on the interpretation of details. Our seventh chapter views the body from a slightly different perspective, examining the ways in which intentional modifications and interventions, such as tattoos, surgical implants and skeletal modifications, affect identification and identity ascription. Our concluding chapter will draw out the key themes and trends associated with the identification and identity surrounding our bodies, and therefore, ourselves.

Clearly the potential scope of this book is huge and it is important that we are clear regarding the perspective that we adopt here for our interpretations. This text aims to serve as a companion reader to those who study the body, but in particular forensic and bioanthropological/archaeological specialists and students. As such, the book presents a discussion of the social constructs of our physical beings from the viewpoint of these disciplines. Specific themes within the context of human identity are discussed in relation to a variety of biological entities of the human body, but those selected and presented here should by no means be considered exhaustive. Each chapter provides an overview of these areas, and we direct the reader to more detailed publications on specific topics where appropriate. It would be impossible to cover every topic in great detail, and that is not our intent. Rather, we seek to present an overarching integrated view on these topics in order to facilitate a more holistic discussion of the body. Our discussions aim to deconstruct and contextualise biological observations regularly made and relied upon in the fields of anthropology and archaeology and to provide a source of discourse that can be applied to other aspects of human body identification. While we acknowledge the significance of the social constructionist perspective on the body and that there is a constructed scientific discourse and practice that seeks a 'truth' about the physiological body as a fleshy entity, we concur with Frank, who states that: 'Empirical bodies do have real limits … corporeality is an obdurate fact' (1991: 49).

2

Categories of identity and identification

[W]e should not be accepting our body as a given, as natural, as pre-discursive, or prior to culture. The body is not a foundation. It is not a biological bedrock upon which we can construct theories of gender, sexuality, race and disability.

(Cream, 1994: 2)

The social descriptors of gender, age, ethnicity and class are pre-eminent in the human identification and identity literature. For the purposes of discourse and analysis, individual and group identity may be conveniently fractured into these four key categories, but in actuality these multiple dimensions act in concert so that, for example, the experience of gender within any one society will also depend on social class, ethnicity and life course stage. While dominant in Western scientific and social discourse, these categories are not meant to be an exhaustive taxonomic representation of human experience; in everyday life a whole spectrum of other identities will come into play (Meskell & Preucel, 2004). Categories serve to construct notions of sameness and difference, and these may or may not reflect the social reality. Categories have boundaries and at boundaries lie discontinuities and tensions. What complicates any analysis of this area is that any individual has access to multiple self-categorisations (Emler, 2005). Identity is also experienced relationally; it is not simply how we view ourselves, but how others perceive us, and thus is much more fluid than the boundedness implied by Western categorisations. Identity can be passively ascribed and experienced, but is often actively constructed and reinforced by both individual agency and society-specific social structures (Díaz-Andreu & Lucy, 2005). Despite these caveats, age, gender, ethnicity and class have profound resonance for how we identify ourselves and others and thus feature extensively in biological and

social science research. It is the dialectical relationship between these different aspects of identity and the physical body that is of interest in this book.
We seek to examine and discuss some of the multitude of ways in which these
aspects of identity become literally embodied by different physiological tissues.
We also seek to understand the ways in which our construction of these categories impacts our scientific analyses of the body. In order to avoid repetition
in later chapters, here we outline some of the current theories and interpretations relating to each of these forms of embodied identity. We briefly discuss
the way in which each of these categories has been conceptualised in the social
sciences and, more often than not, under-theorised in the biological sciences.
We acknowledge that additional forms of identity, such as disability, sexuality
and religion, are just as significant to many, but to examine all of these in detail
is beyond the scope of this book. Our viewpoint throughout this chapter, as in
subsequent ones, is that scientific understandings and constructions of the biological body are not politically neutral; that the cultural constructs of sex, age,
race and class all impact on individual physiology and that these effects in turn
contribute to and reinforce notions of identity.

2.1   Sex, gender, the body and science

> Turn outward the woman's, turn inward, so to speak, and fold double
> the man's, and you will find the same in both in every respect. (Galen
> circa AD 130–200, cited in Laqueur, 1990: 25)

And now we ask a simple question: is the individual a male or a female?
Biological sex, the physiological differences (as perceived in modern Western
consciousness) between male and female bodies, is in many cases considered
the most fundamental aspect of human identification and identity. For many,
sex is a biologically resounding fact. As Freud observed: 'When you meet a
human being the first distinction you make is "male or female", and you are
accustomed to making the distinction with unhesitating certainty' (cited in
Oudshoorn, 1994: 1). Whether one is male or female is proudly announced at
the very moment of birth, or indeed earlier now, with the use of ultrasound
technology, when expectant parents are asked, 'do you want to know the
infant's gender?' (highlighting the confusion surrounding the term 'gender' in
biomedical discourse). One's sex has enormous social ramifications; it governs
every aspect of life, including dress, speech, comportment, occupation, activities and the myriad ways in which humans interact. Here we outline the feminist critique of the category of sex and the influence that gender studies have
had on our understanding and scientific interpretations of the human form.

With respect to human identification, the assignation of sex is perceived as a straightforward process, given the self-evident nature of these bodily differences as alluded to by Freud. The distinctiveness of men and women is strongly emphasised in current biomedical and popular literature: *Men are from Mars, Women are from Venus* (Gray, 1993). On all biological and behavioural levels, males and females have been polarised, and Western society has accentuated this through strongly gendered material culture and behavioural norms. With respect to human identification contexts, a variety of biological measures of difference are used to distinguish between male and female bodies, using techniques perceived as objective and universal. A number of feminist critiques of science have discussed the way in which scientific knowledge has been constructed to reinforce the polarisation of the male–female dichotomy and to play down the overlap between the sexes (e.g. Oudshoorn, 1994; Spannier, 1995; Schiebinger, 2004). As Epstein states:

> Sex differences, like all differences in nature, lie on a continuum, and they become evident through statistical aggregation: there is no unambiguous dividing line between the two sexes, and every criterion of differentiation that might be invoked, from genitalia to hormones to chromosomes, fails to perform a strict demarcating function. (2004: 192)

This biological continuum is strikingly apparent in the authors' own disciplines of bioarchaeology and forensic anthropology, where skeletal variation between males and females is assessed on a sliding scale from hyperfeminine to hypermasculine, rather than as two distinct biological forms (see Chapter 5). Another example is provided by Oudshoorn, who presents a fascinating discussion of the discovery of hormones, describing the way in which endocrinologists came to assign sex to facets of the body that have no inherent sex or gender (1994). The so-called sex hormones are not specific to either males or females, but are contained within the bodies of both and interconverted within them. Males at 60 years of age may have higher levels of oestrogen than females (Spannier, 1995). Oudshoorn also draws attention to a period of uncertainty in the 1920s and '30s when scientists were less sure of their assumptions regarding sex and the body and the clear-cut nature of the differences between the sexes (1994).

Critiques of sex and gender have pointed to the way in which cultural differences between males and females have been naturalised (e.g. girls are less mathematically able than boys). Perceived behavioural differences between the sexes were, in the past, also thought of as purely biological in nature: since the mid eighteenth century, Western society has constructed males as rational, mindful beings, while conceptualising females as irrational and emotional, governed by their leaking, unstable bodies (Sheldon, 2002). Within Western consciousness, women have long been more intimately linked with their bodies, while males have been thought of as more cerebral (Sheldon, 2002; though see Morgan, 2002, for a discussion of male embodiment). As Witz puts it, women over the last few hundred years have been 'overwhelmingly corporealised' (2000: 2). For example, in the eighteenth and nineteenth centuries, when scientists compared women across cultures, they focussed on sexual traits and ethnocentric notions of feminine beauty, such as redness of lips and breast shape. As persists today, the white male body was constructed as the norm and ideal, while females were seen to deviate from this (Tarvis, 1992). Modern interpretations of the human form retain remnants of this thought; for example, the morphology of the female pelvis is often described as a compromise between the optimum biomechanical design for bipedalism – epitomised by the male skeleton – and the need to give birth to a large-brained infant (see Stone and Walrath [2006] for a critique of this). Within this schema, the middle-class white male has no sex, race or gender, but is a generic person (Sheldon, 2002). In another context, Crawshaw has noted similar undertones in men's health magazines, whereby the male body is universal and unproblematic (2007). That this notion persists today is also illustrated by a 1990 report criticising the National Institutes of Health for excluding women from most studies of drug effects, diseases and treatments – as though females are a variable scientists need to control for (Krieger 2003).

Within the discipline of bioanthropology, measurements of the European female skull and pelvis in the nineteenth century led many scientists at the time to conclude that women were the evolutionary inferior of males (Epstein, 2004: 190), mirroring contemporary androcentric views. Within the medical sphere, Epstein discusses the way in which different responses observed between the sexes in clinical studies require no further elaboration; such responses are conceived as arising solely from the biological differences between males and females: cultural context need not apply (2004). Krieger discusses a range of examples drawn from the medical literature whereby 'not only can gender relationship influence expression – and interpretation – of biological traits, but also sex-linked biological characteristics can, in some cases, contribute or amplify gender differentials in health' (2003). Crawshaw also reports how health is an aspect of gendered identity, but largely from the male perspective (2007). Thus the biological and behavioural differences emphasised by material culture and compounded by structured inequalities become naturalised as biological.

The feminist critique of the association between biology, concomitant social roles and inequalities led in the 1970s to the introduction of the word 'gender' (borrowed from psychology) as a term for use in social analysis (Krieger, 2003).

Gender, as defined by Jackson and Scott, 'denotes a hierarchical division between women and men embedded in both social institutions and social practices. Gender is thus a social structural phenomenon but is also produced, negotiated and sustained at the level of everyday interaction' (2002: 1–2). As Simone de Beauvoir famously stated, 'women are not born, but only made' (1974: 301). In other words, sex remains natural, bodily and ahistorical, while gender is a product of culture and history (Chambers, 2007). Within this schema, sex became conceptualised as a fixed dichotomous, biological reality which could be 'read' directly off the body, whereas gender was a more fluid cultural interpretation of these biological differences (Oakley, 1972).

This distinction between sex and gender has been of tremendous importance in the social sciences and in the first instance allowed feminists to argue that 'biological facts' did not explain or justify inequalities and the division of labour between men and women. It allowed researchers to think about masculinity and femininity, not as biological givens, but as historical and cultural constructions rooted in society (Budgeon, 2003). As a consequence of this division, sex became the domain of the scientists, while gender belonged to the social sciences; as a result the 'natural' condition of the human body was left unchallenged (Oudshoorn, 1994: 2).

This science–social theory divide in sex/gender studies becomes apparent when one examines the biomedical literature and the field of human identification, where sex and gender are often confused and conflated (see Walker & Cook, 1998; Krieger, 2003; and Geller, 2005, 2008 for a discussion). Generally this has stemmed from a lack of understanding of the theoretical underpinnings of the use of the word 'gender' within much of the scientific literature, where its usage appears to be confused with a desire for political correctness. Within the discipline of physical anthropology, from which much human identification literature stems, the concept of gender did not feature significantly until the late 1990s onwards, and even then, the majority of this early research was framed primarily in terms of dichotomous biological sex. This situation is now changing and over the last few years a number of researchers have explored the embodiment of gendered practice (e.g. Sofaer, 2006; see Chapter 5 for a discussion). The concept of gender is, however, still almost completely absent (or confused) in a number of other biological disciplines relating to human identification, such as forensic anthropology.

More recent research within the social sciences has problematised the sex–gender binary. In a critique of sex and gender, Delphy stated: 'We have continued to think of gender in terms of sex: to see it as a social dichotomy determined by a natural dichotomy. We now see gender as the content with sex as the container' (1993: 2). The distinction between sex and gender and the rendering of sex as a biological, immutable truth that demands no further investigation has been challenged (Evans, 2002). Sex is now perceived by a number of researchers as as much of a construction as gender (Butler, 1990, 1993; Bordo, 1993), and some have suggested a return to using the term 'sex'. Further, some have reversed the sex–gender binary to argue that sex is a product of gender rather than vice versa. Thus the body has been described as produced within social discourse rather than existing outside it (Schiebinger, 1986, 2004; Laqueur, 1990; Oudshoorn, 1994; Peterson, 1998). As Fournier discusses, 'From this perspective, the sexed body is produced through various gendered mechanisms, or regulatory practices which normalize and mark bodies as male or female' (2002: 57) (see Geller, 2008 and Chapter 5 for a discussion of this in relation to skeletal sex).

Laqueur discussed the historical transience and cultural specificity of the human body, particularly with regard to sex, in detail in his influential book *Making Sex: Body and Gender from the Greeks to Freud* (1990). He demonstrated that not until the eighteenth century did anatomists begin to demarcate and polarise male and female bodies (see also Schiebinger, 1986). Galen, in the second century AD, stated that women were essentially men in whom a lack of vital heat had resulted in the sexual structures being inside the body rather than outside. Laqueur suggests that, until the eighteenth century, a one-sex model prevailed in which anatomical artists tended to use males or females interchangeably to represent the human body (1990). After this time, however, organs that had previously shared a name were linguistically distinguished: 'sex before the seventeenth century was a sociological and not an ontological category' (Laqueur, 1990: 8). Schiebinger argues similarly in relation to the human skeleton (1986). These authors do not claim that there were no perceived biological differences between males and females prior to this time, but that the broader social context influences our understanding of these differences. Of particular significance is that there were no scientific advances in knowledge of the human form at the time of this shift from the one- to two-sex models, underscoring the cultural influence on anatomical interpretation. Sexual difference and the biological facts that describe it are produced through a gendered understanding of the world. For example, Schiebinger suggests that the emphasis on anatomical difference between the sexes from the eighteenth century onwards was a means by which to 'prescribe very different roles for men and women in the social hierarchy' (1986: 43). The work of Laqueur (1990) and Schiebinger (1986) is not without its detractors, including Park and Nye (1991) and Stolberg (2003), who argue that this emphasis on sexual dimorphism within anatomical texts occurred much earlier and was already in place by the sixteenth century.

Existing gendered understandings will always and inevitably inform any account of sexual difference (Sheldon, 2002). Likewise, Butler argues, when describing differences, the observer will always impose his or her own culturally situated understandings and values (1990).

Fournier (2002) and Witz (2000: 2) argue that feminists have been reluctant to engage with discourses on the body because females have for so long been 'undersocialised and overwhelmingly corporealised'. Parallels with this resistance can be similarly observed in studies of disability or impairment, where individuals have been defined almost exclusively in relation to their bodies' perceived dysfunction (see Shakespeare & Watson, 2002 for a discussion), thus spawning a reluctance to return to body-centred discourse. Fournier critiques Butler's stance, stating that 'the body that turns up in this work on body inscriptions rarely turns out to be sentient. It is a body that seems to act as the passive recipient or bearer of inscriptions' (2002: 57). For some, to suggest that the body is constructed through language and culture is to suggest that it has no materiality. This is not to say that language is responsible for bringing the sexes and the body into existence, rather that social context and its linguistic manifestation influence the knowledge we construct about the body (Spannier, 1995; Thomas, 2007: 214). While language is certainly significant for our conceptualisation of the body, flesh and blood is not infinitely malleable and the corporeal raw materials must be given due consideration too.

A number of authors, including Jackson and Scott, argue for the retention of the sex–gender binary, stating that 'we need to challenge assumptions that bind anatomy into gender and sexuality' (2002). Sofaer also argues for the importance of maintaining rather than collapsing the sex–gender distinction in archaeological interpretations of the body. She believes that to do otherwise would create the risk of 'falling back into biological determinism, or of cutting ourselves off completely from the possibility of accessing the full range of potential ways that differences between bodies may be socially regulated and understood' (Sofaer, 2006: 99; for a discussion of sex and gender in archaeology see Gilchrist, 1999 and Sørensen, 2000). The bioarchaeological literature differs from some of the interpretive sociologies in that it generally maintains the sex–gender distinction. Most likely this is because it is a discipline that deals in a very hands-on way with the physical reality of bodies, and male–female differentiation is an important component of this work. Within the 'harder' end of the human identification sciences (e.g. forensic anthropology, pathology, biomedicine), there has been little or no acknowledgement of the separation of sex and gender, let alone any critique of this. 'Doing gender' is not the exclusive business of the social sciences. An understanding of the role of society in shaping the bodies of males and females differently, from the molecular to the macro level, as well as a sensitivity towards the influence of political context in the construction of scientific knowledge, can only serve to improve the quality of our data. Likewise, biological anthropology, whether conducted within the forensic or archaeological sphere, provides a crucial window into the variability of gendered physiologies in relation to a vast array of different cultural practices and beliefs about sex and the body that can inform contemporary debates and disciplines.

2.2   Ageing, the perception of age and the body

'Time is inscribed indelibly on our bodies' (Turner, 1995: 254).

The treatment of age as little more than a 'variable' persisted long after the deconstruction of other facets of identity, such as ethnicity and gender (Sofaer Derevenski, 1997). This unshakeable perception of age as a uniform human experience has persisted because age identity in Western society is associated with (and built upon) the inevitable linearity of time itself (Östör, 1984). Human ageing has been perceived as the physical embodiment of the unidirectional procession of time, set against a backdrop of monolithic cultural behaviour. Time itself is, however, not universally recognised as linear, but subject to variable cultural perceptions and experience. The relationship between time and age in Western society, although linguistically linked and characterised as linear, in reality exhibits an ambiguity that manifests itself through such cyclical associations as childhood and senescence (Fortes, 1984; Gowland, 2006). Indeed, there is some popular recognition that time itself is perceived differently at different life course stages; time is perceived to 'speed up' as we get older. Turner has argued that sociology neglected ageing because it neglected the body (1995). By contrast, within the biomedical sciences and biological anthropology, age has been the focus of a great deal of debate in terms of the mechanisms and processes that result in physiological manifestations of age-related change. Within bioarchaeology, an enormous amount of work has focussed on techniques of estimating age-at-death from skeletal remains (see Chapter 5). The tissues of the body materialise time passed; bones, for example, accumulate osteon fragments, a physiological marker of the passing of time, the body's version of notches on a stick. While the variability of the ageing process among individuals and populations has been a focus of study in this research, the cultural context in which individuals age and the variable social meanings ascribed to the life course have not featured prominently in biological/forensic anthropology or the harder sciences.

Individuals grow up and grow old within social contexts (Campbell & Alwin, 1996: 34), thus age cannot be reduced to physicality or the 'simple passage of time' (Fry, 1996: 117). The study of age identity often utilises a 'life course' rather than a 'life cycle' perspective. Rather than focussing on a series of demarcated age groups, this approach concentrates instead on life pathways and the roles that occur throughout the trajectory of life in a more holistic manner (Marshall, 1996; Moen, 1996: 181). A life course perspective has been described as an 'explicit attempt to view the individual biography within the context of society and to take a historical perspective on both the individual and society' (Marshall, 1996: 22). It is useful to consider age from a biographical perspective rather than a moment in time, because humans are somewhat unusual in that we 'remember our past and worry about the future' (Crews, 2003: 1).

Over the last few decades, research in anthropology, archaeology, social history and sociology has begun to explore the role of age in social identity and as a structuring force in society. Different cultures invariably divide the life course into a series of stages, each accompanied by certain social attributes or expectations of behaviour regarded as appropriate for that age group (Schildkrout, 1978). A biological emphasis on the ageing process within the social sciences almost inevitably resulted in the naturalisation of those social and ideological norms that accompany age-related behaviour. As with gender, age norms for many years were perceived as the social elaboration of a biological given (Moore, 1994).

Ginn and Arber argue that 'a distinction parallel to that between sex and gender needs to be made in relation to age' (1995: 2). This has led several authors to distinguish between a number of different 'types' of age. These are:

1) Physiological age (representing the physical ageing of the body).
2) Chronological age (corresponding to the amount of time that has passed from the moment of birth).
3) Social age (socially constructed norms concerning appropriate behaviour and attitudes for an age group).

Within the human identification literature, it becomes evident that all three definitions of age are actually used: the biological age is translated into a chronological age and then described further by a social age. These various definitions have often been utilised interchangeably, such as child, adolescent or juvenile. They all mean something different, however, and all are culturally loaded: they do not simply convey to the reader a chronological age, but a whole schema of appropriate social behaviour and attributes derived from a modern Western context (Gowland, 2002, 2006).

We cannot deny the biological nature of the ageing process: humans are conceived, are born, develop and grow, reach maturity, undergo physical and/or mental degeneration and die (Sofaer, 2006). The stages of human life history have evolved over millions of years in response to biocultural processes (Crews, 2003). It is well attested, both ethnographically and historically, that age-related social transitions within societies often coincide with physiological parameters (e.g. learning to walk, puberty). This biological framework has undoubtedly resulted in a degree of cross-cultural uniformity with respect to particular social age transitions (Schildkrout, 1978). There is no absolute universality, however, and a biological milestone viewed as important for age categorisation in one society may be entirely disregarded in another (La Fontaine, 1986). As an example, ethnographic evidence indicates that bodily maturity is not necessarily a prerequisite for 'adult' status: thus while the Boran in Ethiopia and Kenya may on occasions recognise a boy's physiological age, others are initiated into 'elder' age sets for non-biological reasons, with subsequent implications for their social identity and relations (Legasse, 1973; Gowland, 2006; Gowland and Redfern, 2010). The physical condition of the body is not necessarily the most important or relevant factor for the conferment of age identity. Even were this so, recent theoretical developments within the social sciences have brought into question the very immutability of these biological transitions.

Cultural practices have a profound effect upon the chronological age of attainment of so-called biological goals. For example, walking and talking is viewed as the beginning of personhood by many cultures, both past and present, and yet cultural practices may either significantly delay or advance such abilities (Levine, 1998; Gowland & Redfern, 2010). Menarche is often viewed as an important physical milestone for females, and yet the age of onset has been found to vary significantly, both among and within populations, due to environmental and socio-economic factors (Beall, 1984). In South Africa, for example, the average age at menarche for those of good socio-economic status is 13 years, for those living in poor rural conditions it is 14 years, and for bushwomen it is 14 to 16 years of age (Henneberg & Lauw, 1995: 4). The body does not develop and degenerate according to a predetermined genetic clock from conception to death (Gowland, 2006, 2007). Indeed senescence, which refers to progressive bodily degeneration once physical maturity has been achieved, is described as an individual phenotype, just as, for example, height and weight are. As Crews states, there is a lack of data that shows any specific genetic programme for senescence (2003: 6). Instead, the body develops according to external as well as internal stimuli and the bodily tissues come to literally embody life course biographies (Robb, 2002; Sofaer, 2006). As biomolecular studies are showing, our bodily tissues are constituted not just by our own biographies but by those of our ancestors (see Chapter 6); through our DNA they create a life course thread that transcends our own embodied identity.

We cannot, therefore, reduce the idea of age to a binary opposition of biological versus cultural age. As Henrietta Moore has stated: 'A more contemporary view of human biology would stress that biology enables culture, while culture brings about biological change ... biology and culture are in a dialectical relationship' (Moore, 1994: 19).

As with all facets of identity, age identity influences and is itself influenced by other social constructs (e.g. gender and status) (Bury, 1995; Bradley, 1996). For example, cross-culturally, the gendered identity of the very old or very young is frequently different from that of adults, with these age groups commonly occupying an almost liminal gender zone. Many anthropological examples exist that demonstrate the culturally androgynous state of older women, and this is often accompanied by an increase in social status and power (e.g. Rasmussen, 2000; Gowland, 2002, 2007). Numerous authors have suggested that this identity shift is related to the cessation of their reproductive role. It does not always coincide with biological changes, however, and tends more often to be connected to social factors only indirectly related to age, such as the marriage of a child or widowhood (Rasmussen, 1987). The same holds true for males who assume roles in later life that are more intimately connected with the domestic sphere. Social ageing is gendered and men and women experience different chronologies (Arber & Ginn, 1995; Moen, 1996). While gender affects the timing of age-related transitions, entry into a particular age group may conversely alter the gendered state of an individual. The interrelationship between gender and age is so striking in virtually all societies that it becomes impossible to examine one of these facets of identity in isolation: the two act simultaneously in the construction of identity (Sofaer Derevenski, 1997; Gowland, 2006; Sofaer, 2006).

Hockey and Draper have sought to examine the boundaries of the embodied life course (2005). They demonstrate that the choice of bodily indicators that 'life' has started or stopped is highly contingent, depending upon political position, historical location and available technologies. For example, pregnancy tests that can identify conception at increasingly early stages are resulting in the identification of many more early miscarriages – a category that did not feature prior to this technology. Indeed, embodied practices of would-be mothers (vitamin supplements, fitness regime) even serve to formulate a pre-conception identity of a hoped-for child (Hockey & Draper, 2005). On the other end of the spectrum, the identification of the moment of death is becoming less absolute with increasing medical advancements, a difficulty exacerbated by the practice of organ donation (see Chapter 4). As archaeologists are well aware, even long-dead individuals may exert a powerful agency and identities may be ascribed to individuals by others across a temporal disconnect of many millennia (e.g. see Peers, 2009; Robb, 2009). Hockey and Draper note that 'identification is therefore a negotiated process, tied to the body – but not limited to the body … A model of the life course which accommodates these moments enables a more expansive conceptualisation of social identity which recognises its nature as a relational, inevitably incomplete social process' (2005: 48).

Technological advances hold very profound implications for our conceptions and lived experience of the life course. Brown and Webster have argued that 'the lifecourse is being reshaped, opened up temporally and spatially, not only in an attempt to address physical deterioration but also to redefine the boundaries of life itself (as in stem cell research) … Just as importantly, this reshaping also produces new entities whose moral bearing is highly contested [e.g. early embryos]' (2004: 75). The fundamental uni-linearity of embodied time is challenged by these new technologies, for example by the introduction of progenitor stem cells into the human body (Brown & Webster, 2004). Ever increasing numbers of people now survive beyond the age of 70 years and this presents an emerging area of medical and biocultural research (Crews, 2003). The impact of these demographic shifts is usually discussed in negative socio-economic terms, but the shifting of the upper boundaries of the life course will no doubt ripple down, leading to a renegotiation of younger age identities.

Theoretical developments within the social sciences have led to an understanding of the complexity and fluidity of age identity. Ethnographic and historical examples reveal that biological, cultural and chronological concepts of age are intertwined in the formation of age identity in a way that is impossible to unravel. This complexity is further exacerbated by the interrelationships among age, gender and status. As discussed, gender is biographically located, while age transitions are gender specific. These biographies are enmeshed within the different tissues of the body so that our physicality has implications for our age identity. Human identification scientists seek to extract meaning from the different layers of the body in an attempt to reconstruct aspects of our biographies. In so doing, they must be sensitive to the culturally contingent nature of the ageing process.

2.3   Race, ethnicity and the body

Racial categories have formed a key basis for human identification for many centuries and, despite the more recent dismantling of the biological basis for such categorisations, they remain a powerful means of grouping people today. Such is the potency of racial categorisations that one's 'ethnicity' has been shown to have a dramatic impact on health and socio-economic well-being, influencing as it does everyday interactions and the attainment and shaping of social aspirations (Gravlee, 2009). In the United States, the category of 'race' has been used since the first census in 1790 (Nazroo & Williams, 2006: 240). In today's bureaucratic world, declarations of ethnic identity are a common request: we assign ourselves to one ethnic grouping from a choice of seemingly idiosyncratic descriptive categories which fall under the ethnic banner (though in reality these are based on a smorgasbord of skin colour, religion, linguistic commonalities and other non-hereditary factors). This apparent intangibility is, however, contrary to the biological resonance that 'race' has in everyday life. In most societies, cultural constructions of race are perceived as far from ephemeral; instead they exert a powerful and often malign influence on the lives of individuals and groups. The construct of 'race' therefore forms a key component of human identity and so features prominently in the human identification literature.

According to historians, the notion of race in the Western world has existed only since the late sixteenth century, following the great sea explorations of that era (Brace, 1995). The first known reference to the word 'race' appears in Tant's *Thesor de la Langue Francaise* of 1606 (Lieberman, 1975). From the eighteenth century onwards, scientists began systematically constructing race as bodily difference by grouping humans together and apart based on physical characteristics (Lieberman, 1975), on both a local and national level. In the eighteenth century, Linnaeus introduced the idea of four varieties of humanity to the scientific literature in his 1735 *Systema Naturae* (Lieberman & Reynolds, 1978: 333–4). Linnaeus saw the human species as a fixed and unchanging entity made up of four races distinguished primarily by skin colour – a powerful means of categorising individuals to this day (see Chapter 3). Other significant physiological descriptors for categorising races in the past included craniometrics (see Chapter 5), hair texture, lip colour and genitalia. Numerous categorisations of the human form were developed after Linnaeus' schema, with one of the most enduring being Blumenbach's five-race division based primarily on skull form (Caucasian, Mongolian, Ethiopian, Malayan, American). The discipline of physical anthropology has its roots in the classification and description of the variation in the human form, and early anthropologists devoted much of their work to cranial measurements. Racial categorisation was a taxonomic exercise conducted within an academic environment heavily focussed on typologies of the natural world (Abu El-Haj, 2007).

These categorisations, however, did not serve only to passively describe human physiological variation, but instead became associated with and served to reinforce the ideology of racism (Blakey, 1987). This is the conviction that race and behaviour are linked to heredity and that some races are superior to others (Lieberman, 1975). Biological differences in themselves have no intrinsic social meaning (Nelkin & Lindee, 1995); such differences, however, came to be constructed as though they reflected more fundamental differences among humans, in particular, intellectual capacity and behaviour (Keita & Kittles, 1997; Gravlee, 2009). Early racial typologies helped to establish and legitimise the ideological framework of this time that proclaimed the biological and social superiority of the white male (see Gould, 1997; Mukhopadhyay & Moses, 1997; Ahmed, 2002). Such a viewpoint was highly conducive to an environment of colonialism and slavery (Gould, 1997). As Ahmed notes, 'the invention of race as something that belongs to bodies, and belongs to different bodies differently, was a means of justifying and legitimating the period of imperial expansion in the 19th century' (2002: 46). The inequality of the races remained the predominant view of scientists and intellectuals until the early twentieth century. Categorisations of sex and class cut across those of race in such a way that women and the poor were considered much lower on the evolutionary ladder than white males. A view 'confirmed' by early craniometrists whose findings of small cranial capacity (measurements that did not take into account stature) demonstrated that blacks, women and the poor stood at the bottom of the evolutionary ladder (Gould, 1997; Epstein, 2004).

As Blakey discusses, the body became the site through which social inequalities became naturalised and thus were a biological issue rather than one of government policy (1987). An ensuing progression to this way of thinking was that action must be taken to prevent a physiological superior nation being 'watered down' by the blood of inferior human beings. Thus, eugenics came into being. 'Eugenics' was a term coined in 1883 by Francis Galton (cousin to Charles Darwin), who believed that marriage and the number of offspring should be regulated according to parental 'fitness' (Gould, 1997). While eugenics in the United Kingdom was associated with class prejudice, in the United States and other countries (e.g. Germany) it was more explicitly directed towards race. Although generally abhorrent to us today and widely associated with the horrors of Nazi genocide, up until the post-war period eugenic thought was widely supported by scientists and politicians throughout the Western world and beyond and had a profound influence on public policy at the time. For example, sterilisations were performed in mental health facilities on men and women who were considered feeble minded. Such practices were broadly accepted as a rational approach to population control. Franz Boas was one of the few to argue against eugenic practices, emphasising the role of the social environment in the development of 'desirable' traits (Caspari, 2009). In the post-war period, the United Nations mitigated against the espousal of eugenics and racial hierarchies (Blakey, 1987: 25). At this time, the horror of the Holocaust and the abuse of overly simplistic science in order to reach 'social destructive ends' was realised (Allen, 1997: 87). That said, there has been a resurging interest in the subject of eugenics in recent years in relation to genomic research. This has sparked some interesting debates, for example in relation to prenatal genetic screening for abnormalities – where does one draw the line? (see Chapter 6).

Early critics of the race concept were Franz Boas and his students, who challenged the notion of race as a group of biologically fixed traits. Boas criticised the models of racial analysis used by his contemporaries in anthropology which served to reinforce the biological rootedness of social inequalities (see Blakey, 1987; Mukhopadhyay & Moses, 1997). As early as 1894, Boas explicitly rejected racial determinism of culture. He was interested in the effect of environment on the physical form and was a pioneer in the study of human biocultural plasticity (Caspari, 2009). In 1912, he published his now famous study of nearly eighteen thousand Southern and Eastern European immigrants in New York, *Changes in Bodily Form of Descendents of Immigrants* (1912), for the Senate Immigration Commission. In it he emphasised the plasticity of the human form by demonstrating the changes in the cranial indices of children of first-generation immigrants to the United States (Blakey, 1987; see Gravlee et al., 2003 for a more recent reanalysis of Boas's data). Some authors have argued that because of Boas's emphasis on the malleability of such indices, he was a pioneer in the now burgeoning field of embodiment (Gravlee et al., 2003; Gravlee, 2009). Boas was instrumental in an American Anthropological Association resolution against 'scientific racism' in 1938 (Blakey, 1987: 23). In 1935, Huxley and Haddon wrote, 'In the circumstances, it is very desirable that the term *race* as applied to human groups should be dropped from the vocabulary of science' (cited in Billinger, 2007: 13). By the end of the Second World War, social and environmental perspectives on the variation in biology and health largely predominated and nineteenth-century racial categorisations looked less fixed. Instead, the work of anthropologists such as Montagu and Brace in the 1960s argued for the socially and culturally constructed nature of race. With the advent of biomolecular techniques of analysis, the edifice of discrete, fixed, biological racial categories was fundamentally shaken. Throughout the course of the twentieth century, work on blood types and DNA evidence helped to establish that the range of genetic diversity within groups far exceeded that observed among groups. At this time, there was a shift from the use of 'race' to 'population' in describing human groups and 'typological thinking was replaced by statistical thinking' (Abu El-Haj, 2007: 286). Population genetics created a vision of dynamic populations with overlapping gene frequencies (Mukhopadhyay & Moses, 1997), leading to the much quoted phrase by Livingstone: 'there are no populations, only clines' (1962: 279). The notion that humans can be naturally divided into a few discrete biological subdivisions was effectively dismantled (Keita & Kittles, 1997; Smedley, 2007). By the 1980s, anthropology, the discipline founded on the construction of racial difference, seemed to have successfully challenged the existence of a biological basis in racial categorisations (Jablonski, 2004):

> In the end, the terms, 'race', 'ethnicity' and 'ancestry' all describe just a small part of the complex web of biological and social connections that link individuals and groups to each other. (Race, Ethnicity and Genetics Working Group, 2005: 524)

'Ethnicity', which emphasises the cultural, socio-economic, religious and political qualities of human groups rather than phenotype or genetic ancestry, has largely replaced 'race' in academic and bureaucratic parlance. Montagu argued that 'ethnicity' represents a new way of conceptualising human variation, because of the demonstrable effect of social environment on evolutionary patterns. Billinger suggests that the use of the term 'ethnicity' instead of 'race' served to 'neutralise the territory' and negate the violence of the history of racial and racist discourse (2007). This was by no means the end of race as a social category within society (Mukhopadhyay & Moses, 1997). While it has been established that racial categorisations as used today have no biological basis, in everyday life the concept of race 'retains an aura of self-evident naturalness and … a profound political salience' (Epstein, 2004: 195). Within the forensic anthropological setting, some still consider race one of the 'four pillars' of osteo-profiling, and race is one of the pieces of information investigating officers expect from the anthropologist. In their review of the subject, Sauer and Wankmiller note that, despite rejection of the concept of race by many forensic anthropologists, some have strongly argued that to not provide such an assessment is in fact more unethical since in the medico-legal context the 'traditional schemes of race have clear meaning' (2009: 196). In the United Kingdom, many anthropologists have abandoned the use of the race concept in teaching physical anthropology due to the arbitrary nature of many of the phenotypic divisions used and the poor correlation between traits and geographical location (Armelagos 1994). After a long period of abandonment, craniometric studies are, however, once more being used on past and present populations to examine, not explicitly race, but population affiliation (see Chapter 5). What anthropologists are describing are phenotypic differences, but the link between phenotype and genotype is far from straightforward. Phenotypic variables used for racial categorisations are in reality extremely plastic and strongly influenced by factors other than genetics (see Chapter 5). The use of DNA analysis for the analysis of 'ancestry' is widespread within human identification contexts

(Sarich & Miele, 2004: 22). Kennedy has highlighted the fact that there is a two-tier system in place whereby any biomolecular studies of population affinity are tolerated, whereas studies of morphological traits are seen as antiquated and even racist (1995: 798).

Contemporary beliefs about inner racial differences persist, perpetuating current inequalities (Hirschfield, 1996). After all, *racism* is not dependent upon scientific typological categorisations of difference (Abu El-Haj, 2007). For example, in a study of whites who placed online dating adverts, 50 per cent said that race was not important, but 90 per cent of those individuals replied only to white respondents (Hitsch et al., 2004). Further, this perception of race has a considerable impact on the body because it creates an 'unequal structuring of life chance' (Mullings & Schulz, 2006: 3), leading to profound health inequalities. Therefore, while 'race' is a socially constructed phenomenon, it functions as a significant analytical category because of its social salience and subsequent effect on people's lives (Kuzawa & Sweet, 2009). Epidemiological studies frequently employ 'race' or 'ethnicity' as a key variable. For example, Comstock et al. reported that 77 per cent of articles published between 1996 and 1999 in the *American Journal of Epidemiology* and the *American Journal of Public Health* made reference to either race or ethnicity (2004). Yet such research generally does not define or justify the racial categorisations employed or sufficiently consider cultural factors when interpreting evidence (Dressler et al., 2005). Shim states that researchers often use race as 'an imperfect proxy' for cultural differences (2005: 414). The utility and meaning of racial categories in biomedical research is heavily debated (Abu El-Haj, 2007). Recent studies have adopted a much more nuanced approach to the relationship between health and ethnicity. For example, Gravlee et al. (2005) have demonstrated an association between skin colour (interestingly, self-rated or ascribed rather than actual pigmentation) and blood pressure within some ethnic groups due to the interaction with income, education and psychosocial stressors (Gravlee & Dressler 2005). In previous medical discourses, racial disparities in health were ascribed to genetic, biological differences and associated lifestyle choices, rather than the outcome of social relations and inequalities.

Studies have highlighted health disparities between black and white Americans on nearly every index measured (Dressler et al., 2005: 232). Socio-economic status does not entirely explain the health disparities observed, and instead studies now focus on the relationship between psychosocial stresses rooted in racism and health. Gravlee argues that the challenge facing scientists of the body is to move beyond the past assertion that race is not biology to explain how race *becomes* biology (2009). Recent research on racial inequalities in health highlights phenotypic plasticity and a complex biocultural view of human biology (Gravlee, 2009; Kuzawa & Sweet, 2009). As Shim observes in her study of cardiovascular disease and ethnicity:

> Race acts, in concert with other dimensions of inequality, to shape the conditions into which people are born, the opportunities they have throughout their life course, the problems or risks they encounter at different stages of life, and perceptions, attitudes, and treatment by other people and institutions. Inequalities in access to resources such as knowledge, money, power, prestige, and social connections – effected through social relations and structures – influence multiple health outcomes, including cardiovascular ones, and therefore act as fundamental causes of disease. (2005: 431)

As has been pointed out already, 'it is a vicious cycle: social inequalities shape the biology of racialised groups, and embodied inequalities perpetuate a racialised view of human biology' (Gravlee, 2009: 48). In 2004, the overall age-adjusted death rate for black Americans was more than 30 per cent higher than it was for white Americans; for some leading causes of death, the disparity was substantially higher (Gravlee, 2009). It is important to note that not only contemporary inequalities are embodied in racially disadvantaged groups, instead we are observing the embodied impact of cumulative disadvantage across not only an individual's life course but across generations (see Chapter 6 for a discussion of social disadvantage and DNA). In this book we follow Gravlee's view that an analysis of biological differences among 'ethnic' groups is justifiable if one is interpreting these, not solely in terms of genotype, but in relation to sociocultural and environmental interactions (2009).

2.4   Socio-economic status

> Class is always coded through bodily dispositions: the body is the most ubiquitous signifier of class. (Skeggs, 1997: 82)

One's status in society is popularly conceptualised as a more exclusively *social* category than sex, age and ethnicity. Socio-economic status has, however, profound biological consequences: it is the strongest determinant of health variations (Jackson & Williams, 2006: 136). Again, due to the intersectional nature of identity, the experience of other social categories will also vary in any one society depending on social class. As Krieger and Davey Smith observe: 'The construct of embodiment invites us to consider how our bodies accumulate and integrate experiences and exposures structured by diverse yet commingled aspects of social position and inequality' (2004: 99).

Socio-economic status is a key factor determining the biocultural milieu in which humans interact. Education, diet, housing, access to health care, exposure to pollutants, infectious diseases and parasites, physical activities and psychosocial stressors are all related to one's position within the social stratum. Given this, it is unsurprising that a strong correlation exists between social status and physiological well-being (Brunner & Marmot, 2006). Our physiology embodies culturally specific social inequalities.

The relationship between body and wealth was first analysed in detail by Louis Rene Villermé in his research on the risk of mortality, short stature and illness in Paris in the 1820s. Bodily differences were related to wealth differentials and variation in sanitary facilities (Szreter, 2005). Social inequalities in mortality were recorded much earlier than this with some historical evidence from the sixteenth and seventeenth centuries in Britain (Whitehead, 1997). Archaeological studies have sought to examine the relationship between health and wealth from much earlier periods by analysing skeletal indicators of health stress in relation to burial ritual and grave goods (e.g. Craig & Buckberry, 2010 for the Anglo-Saxon period).

Childhood growth and adult stature is still viewed as an important correlate of socio-economic status. A 1958 British birth cohort study found that height at age 7 years was a strong predictor of unemployment in later life, with the chances of unemployment in the shortest fifth of children being three times greater than that of the tallest fifth. The same study indicated that retarded growth at the age of 7 years was likely the consequence of socio-economic uncertainty and psychosocial adversity in the lower status groups; furthermore, parental social class is a predictor of birth weight (Blane, 2006).

Health in early life forms the basis of health in adult life (Wadsworth & Butterworth, 2006). Researchers have attempted to model this relationship amongst past populations by examining the relationship between skeletal indicators of physiological poor health, including growth and adult stature and mortality in archaeological cemetery populations (see Chapter 5). Further, stresses experienced during growth and development are thought to influence the presence or absence of a variety of phenotypic variables (e.g. epigenetic traits; see Chapter 6) along with asymmetry in the body. Status-related impacts on the body are potentially important in terms of human identification, but also for interpreting past living environments in relation to health and identity.

In the nineteenth century, Edwin Chadwick was deeply concerned that the conditions of the working classes resulted in stunted growth and poor health (Chadwick, 1965). Likewise, Engels was horrified by the physical appearance of the working-class people he encountered in nineteenth-century cities, describing in detail a range of bodily deformities (1845). These included deviations in the spinal column, scrofula and rachitic legs, all of which are highly visible. In these instances the inferiority of social status was embodied through inferior physical forms, and this would have served as markers of identity that reinforced the middle-class perception of the poor as being almost subhuman. Such associations would fan the flames of class eugenics in the early twentieth century. The poor physical condition of the working classes in northern England in the nineteenth century was the result of inadequate nutrition, long working hours from childhood onwards and the breakdown of the family unit as a result of all family members toiling in factories, resulting in inadequate infant care. The descriptions supplied by Engels correlate with what recent skeletal analyses have revealed (e.g. at St Martin's-in-the-Bull Ring, Birmingham; Brickley et al., 2006). In Britain, intensive industrialisation during the nineteenth century brought a dramatic discontinuity in population health which lasted from the 1820s to the 1870s. Szreter observed that in the 1830s and 1840s, life expectancy at birth in such cities plummeted to levels not seen since the Black Death in the fourteenth century (2005: 31).

Demographic historians have shown that figures of life expectancy at birth for the upper classes first began to exceed the average for Britain after 1750 (Szreter, 2005). That is not to say that status differences in health were not apparent prior to the eighteenth century. By the mid nineteenth century in Liverpool, the average age of death for labourers was an astonishingly low 15 years. This is less than half that recorded for the gentry in the city (Whitehead, 1997). While no longer so extreme, health inequalities persist to this day. For example, inhabitants of the poorer suburbs of Glasgow were recently reported to have a life expectancy of 12 years less than those living in more affluent areas of the city (NHS Scotland 2004, cited in Marmot, 2006: 1). Even greater disparities exist among different countries as a result of differences in environment and psychosocial stressors (Marmot, 2006).

Inequalities in health and mortality related to social status are currently widening (Graham, 2000). In addition, geographical inequalities now stand at the highest levels ever recorded (Dorling et al., 2000). That inequalities in health exist according to social status is perhaps not surprising, but studies have shown that it is not the *wealth* of a nation per se that is of greatest significance, but the way in which wealth is *distributed* throughout society. As Shaw et al. explain: 'It is not only that the poorest in society have poor health, but a gradient of ill health and mortality spans all socio-economic strata' (2006: 196). Overall life expectancy is generally higher in societies which have greater equality among their citizens (Graham, 2000). For example, the United States has a GDP per capita twice as high as Greece, yet life expectancy is higher in the latter (Wilkinson, 1999).

As mentioned previously, the experience of socio-economic status will vary according to ethnicity/race and sex/gender. For example, a study of infant mortality amongst European Americans and African Americans in the United States demonstrated that at every educational level infants born to black mothers were approximately two times more likely to die than those born to white mothers. The mean life expectancy for African Americans is almost 6 years less than white Americans and these gaps persist (Geiger, 2006). While much of this has been linked to biological and behavioural factors, more recently the role of psychosocial stressors has been investigated (Jackson & Williams, 2006). Indeed, an important change in contemporary understandings of the social determinants of health is the recognition of the impact of psychosocial health factors (Wilkinson, 2006: 341). These have proven particularly significant in terms of health outcomes, and researchers are currently focussing much more on the biological impact of the social environment (Brunner & Marmot, 2006). Related to this is the research on epigenetics (see Chapter 6), which highlights the significance of diet and social environment on the pattern of gene expression. For example, maternal stress can influence the developing foetus at a molecular level such that health outcomes in later life will be affected. Social inequalities modify key biological processes and are thus embodied at the molecular level and perpetuated across generations (Kuzawa & Sweet, 2009). Indeed it has been argued that the low birth weight of African American infants compared to their European American counterparts is a consequence of conditions experienced by the former during generations of slavery; essentially, it is the biomolecular inheritance of poor ancestral environmental conditions that have yet to be successfully mitigated by the contemporary social milieu (Jasienska, 2009).

The significance of psychosocial stressors is illustrated by the fact that income and health are related within developed societies but not among them. Thus it is not only the direct physical effects of exposure to better or worse material conditions that is at issue, but the crucial factor is one's position in the social hierarchy. In particular, stigmatisation and exclusion of those at the very bottom have particularly profound health consequences (Shaw et al., 2006; Wilkinson, 2006). An experimental study which sought to examine the cognitive effect of mild social exclusion using computer-simulated MRI imaging of the brain showed that the feelings of social exclusion activated the same part of the brain as physical pain (Wilkinson, 2006: 345).

Social status is intertwined with other aspects of identity in a complex web. The tissues of the body incorporate the resulting social and biological experiences of being wealthy or poor in relation to these other social variables within a particular society. The literal embodiment of these processes then serves in a mutually constitutive manner to reinforce social stereotypes, regarding, for example, the inferiority of the nineteenth-century working-class person. One may accumulate biological markers of disadvantage throughout one's life course; poor maternal environment, weaning during infancy and diet during childhood may result in poorer health outcomes later in life. Epigenetic studies have demonstrated that individuals do not simply embody the disadvantage they are born into, but they carry forward the weight of inequalities endured by their ancestors. As Blane writes: 'A person's past social experiences become written into the physiology and pathology of their body' (2006: 54). Socio-economic status is therefore a highly significant factor in terms of both human identification and identity studies.

2.5   Conclusion

This chapter has aimed to highlight a range of body-mediated categorisations that have profound significance in everyday life. The following chapters return to these aspects of identity in relation to the different tissues of the body. Some of the key themes drawn out by this review may be summarised as follows. While much research within the social sciences has tended to focus on a single aspect of identity such as 'gender' or 'ethnicity', in actuality it is difficult to tease apart one from the multiplicities of constitutive identities that orchestrate a persona. For example, gender expression is altered according to life course stage, and those of different social classes and ethnicities will construct and experience gender differently in ways which interact with the body as a corporeal as well as a social entity. The biological reductionism of past views of identity has given way in more recent decades to social constructionism, in which the role of discourse in constructing not only social identity but physical understandings of the body has been emphasised. This discourse has been extremely important for challenging our understandings of the immutable nature of identity. The body as flesh and blood, however, was marginalised within this work, and it further served to reinforce the science–social theory divide that has characterised much body-centred research (Shilling, 1993). More recent studies have placed the corporeal body at the centre of identity discourse and have sought to examine the interrelationship between the physicality of the body and the social milieu. The following chapters explore this interrelationship, starting with the largest and most visible organ of the human body, one that lies at the boundary, the coal face, of identity relations and that bears the scars of these interactions: the skin.

3

The skin

There should be more than this flimsy dermal bubble separating the vastness of the cosmos from the throb of blood and consciousness that is you.

(Jones, 2007)

The skin, or integumentary system, is perhaps the most significant region of the body in terms of human identity and identification. In physiological terms, its role is significant since it forms the physical barrier or border between our internal organs and the outside world. In social terms, 'the skin gives us both the shape of the world and our shape in it' (Connor, 2004: 36). It provides a vital conduit between these worlds via the sense of touch, providing the body with a constant stream of information concerning our immediate environment and stimulation, including pleasure and pain. The skin is responsive to emotional stimuli (Picardi & Pasquini, 2007; Morrison et al., 2010). Touch is the first of our senses to develop and skin-to-skin contact between baby and mother is thought to be critical, not only for intimacy and bonding, but also for neuropsychological development (Weiss, 2005). As Benthien discusses, a newborn baby laid against the skin of her mother does not yet know that she is a separate being. Gradually she learns where her boundaries are through tactile experiences intimately tied to her emotional sense of well-being (2002).

The skin should not be viewed as a solid boundary, since it is permeable and allows the interchange of substances between the body and its environment (Connor, 2004). The skin simultaneously communicates multiple facets of identity, on both a 'biological' and cultural level (e.g. ancestry, age, gender, health and external modifications). It is malleable and plastic, changing throughout the life course from birth to old age and in response to environmental and cultural stimuli. Comprising approximately 16 per cent of body weight (Gawkrodger, 2002), the skin appears as the most external and exposed anatomical region and, as such, it is the skin that others first notice and make judgements based upon. The predominantly hairless skin of the human has been likened to a blank canvas which is both passively imprinted by our social and physical environment and actively manipulated or inscribed in order to communicate (or obfuscate) aspects of our cultural identity to and from the world through tattoos, body modifications and cosmetic interventions (Schildkrout, 2004). The skin as the external surface of our body reflects ourselves and the image we wish to project, and we adorn it accordingly. The skin, however, is not passive like a canvas. It can betray us too. Its pallor alters, providing a window to inner health: it becomes blemished and blotched, it stretches, sags and wrinkles; it is a repository for fat – the scourge of the developed world and closely aligned with socio-economic status. Entire industries have been built on products to 'control' this important outer surface of the body through cosmetics and surgery, to mask or manipulate our appearance and perceptions about who we are.

The skin represents the largest organ of the human body and, as a consequence, is well represented in the brain (Montagu, 1978); indeed, the skin and the brain are formed from the same membrane in utero, the ectoderm. The sense of touch provides the world with a material reality that for humans is often perceived as more reliable than sight. We require confirmation of the visual through the tactile (Benthien, 2002). The skin also marks the first historical focus of identification methodologies, discussed as it is in Sung Tz'u's forensic manual *Collected Writings on the Washing Away of Wrongs* of AD 1247. This chapter explores the role of this key, boundary-forming organ in human identification and identity interaction.

3.1   The structure of the skin

The structure of the skin was overlooked by early anatomists, including Galen, who seemed more preoccupied with what lay beneath. This changed in the seventeenth century with Cowper's *The Anatomy of the Humane Body*, in which the anatomical layers of the skin were first described with the assistance of the microscope (Wilson, 1999: 65). The key biological function of the slightly acidic skin is to provide protection for the inner environment of an organism and to thus deflect the myriad physical, chemical and biological insults facing the organism on a daily basis (Escoffier et al., 1989; Jablonski, 2004; Has & Bruckner-Tuderman, 2006; Calleja-Agius et al., 2007). It also plays a vital role in maintaining bodily temperature, integrity and hydration, providing grip and sensory information and producing vitamin D (Gawkrodger, 2002; Calleja-Agius et al., 2007). Additionally, it contains dendritic cells to stimulate immune responses (Giacomoni et al., 2009). As such, each layer of the skin has different mechanical and structural properties (Diridollou et al., 1998). This can be seen in recent research investigating one aspect of this protective property of skin, namely the effect of UVA and UVB radiation from exposure to the sun, which Fisher et al. highlight as the most common source of environmental damage to the skin (2008). Interestingly, the effects are not uniform through the stratigraphy of the skin, with UVA affecting the outer epidermis and UVB the deeper dermis (Vioux-Chagnoleau et al., 2006). Failure of the skin to uphold its general protective duties results in serious and unpleasant responses to microbial infection and allergen exposure (Sicherer & Leung, 2007).

In structure, the skin consists of three basic layers: the epidermis, the dermis and the subcutis. Each of these layers in turn consists of its own individual function and structure. The epidermis, although only 0.1 mm thick, is constructed of a number of different layers: the stratum corneum (a collection of fifteen to twenty layers of flattened dead, but functionally important, cell remnants), the granular, spinous and basal layers and the basement membrane (Cauna, 1954; Champod et al., 2004; Has & Bruckner-Tuderman, 2006). These layers are not discrete features, but instead represent the maturation of keratin (high-molecular weight polypeptide chains) with their production within the deeper basal layer and their final death in the corneum (Gawkrodger, 2002; Has & Bruckner-Tuderman, 2006). Along the way, the cells transform from a columnar shape to a polygonal one and, within the outermost layer, the dead cells thicken to form a strong, flexible casing that can withstand three times its own weight in water. Studies show that cell progression from the basal to corneum layers takes approximately two weeks – the same time period for cell shedding once it has reached the final layer. The integrity of these five layers is possible due to the cohesion and adhesion of the keratinocytes within the epidermis, and the keratin provides considerable protection from mechanical and non-mechanical stresses suffered during life (Has & Bruckner-Tuderman, 2006).

The dermal layer is a more supportive network of connective tissue, and contains the specialised structures of the skin, such as the blood vessels, nerve fibres and sebaceous glands (Gawkrodger, 2002). Delicate nerve fibres and the sweat ducts are the only structures to pass from the dermal layer into the epidermis (Ashbaugh, 1999), with the nerve fibres providing up to thirty times more sensitivity in the fingertip region (Cauna, 1954). The dermis primarily acts as a cushion by bestowing elasticity and plasticity to the skin (Escoffier et al., 1989; Diridollou et al., 1998; Has & Bruckner-Tuderman, 2006), although it also performs an important role in providing nutrients to the epidermal layer (Ashbaugh, 1999). The response of the dermis is affected by the nature of the stresses applied to the skin (Delalleau et al., 2008), although in normal functioning situations the dermis is subjected to continuous anisotropic stresses (Escoffier et al., 1989) with which it copes well. The dermis varies in thickness from between 0.6 mm (eyelids) to 3 mm (back and palmar surfaces) and is also a layered structure containing the papillary dermis (an undulating layer that mirrors the basal layer of the epidermis) and the reticular dermis beneath (Gawkrodger, 2002). Overall skin thickness has been shown to be 16 per cent greater in men compared to women (Escoffier et al., 1989). Seventy per cent of the dermis is collagen, of which Type I is the most abundant, whilst the remainder is composed of loosely arranged elastin fibres, the semi-solid ground substance (a matrix of glycosaminoglycans which allows for mild structural movement), hair follicles, sweat glands, fibroblasts (which form the connective tissues, collagen, elastin and glycosaminoglycans), dermal dendrocytes (which have an immune capacity), macrophages and lymphocytes (Gawkrodger, 2002; Calleja-Agius et al., 2007; Fisher et al., 2008). The collagen fibres are arranged in a formation parallel to the skin's surface, thus providing a high tensile strength to the tissue (Bischoff et al., 2000; Calleja-Agius et al., 2007).

Between the epidermis and the dermis is the dermal–epidermal junction, a vital region for adhesion and signalling between the two main layers of skin (Has & Bruckner-Tuderman, 2006). The performance of this junction has been shown to improve with increased quantities of vitamin C (Vioux-Chagnoleau et al., 2006). The average thickness of the epidermis and dermis ranges between 1 and 1.5 mm (Delalleau et al., 2008). Only damage to the dermal layer of the skin will result in permanent scarring on the covering epidermal layer (Champod et al., 2004). Below the dermis lies the subcutaneous layer, simply composed of loose connective tissues and fats (Gawkrodger, 2002).

3.2   Skin and identification

Without the skin we cannot be recognised. Skin cloaks and helps shape our form and it is through familiarity with our skin's surface that friends and acquaintances can identify us. Without it, identification can only be achieved through impersonal techniques divorced from such social connections. Such is the importance of the skin that even in death great pains are taken by many cultures to preserve its integrity. An essential part of the process of embalming or mummification is the removal of the internal organs: ironically the organs are protected by the skin during life, but threaten it in death (Connor, 2004).

In archaeological contexts, the skin rarely survives except in unusual circumstances (e.g. mummification, bog bodies) and even then the texture and colour is far removed from that of the living. In those rare instances when the skin is preserved, the human remains provoke a much more emotive response – a greater connection is generated between the observer and the dead. Only when the bones are fleshed and the body's living surface is retained do we feel that we are truly face to face with our ancestors. Such burials also provide a wealth of information usually lost to archaeologists, including elaborate tattoos, the presence of diseases such as smallpox (e.g. eighteenth-century crypt burials at Spitalfield, London) or indications of violence (e.g. the 'Lindow Man' bog body who had been garrotted and his throat slit). As a consequence, these burials generate a great deal of public interest, feature in the popular media and are the focus of exhibitions – all because the skin is present. The 'Ice Man' is a particularly well-known example. This is the mummified body of a man who died approximately five thousand years ago and whose frozen body was recently recovered from the Alps. As Robb discusses, it is precisely because the skin and features of this individual were so well preserved, and hence it (he) was so 'compellingly human', that particularly strong feelings regarding the ethics of its study and display were invoked (2009: 109–11). Burials of skeletal remains alone do not garner such attention; a vital part of their humanity is absent as is the unique biographical and social data retained by their skin (referred to by Ahmed and Stacey as 'dermographies' [2001]). The depersonalisation conferred by the lack of skin is highlighted by the Body World exhibitions in which real bodies, flayed of skin, are displayed undertaking a variety of activities. Were these individuals to have their skin intact, one would expect the public response to them to be more repulsion than captivation. Other considerations surround this exhibition, such as the degree to which we are fascinated by the things most will never see in their lifetime – since the skin will prevent this viewing.

Within the forensic sphere, the skin's surface provides a multitude of means of identification, even in the absence of friends or family. The decomposition of the skin has important repercussions for identifying victims or perpetrators, either living or dead. As our skin and its features disintegrate we dissolve into anonymity. We become removed from the realm of the social and enter the world of science: we transform from person to object. But nonetheless the residues of the skin can remain, perhaps in the form of prints on a myriad surfaces touched during a person's existence. Thus the skin is one of the key focuses of the identification sciences, a relationship we explore further throughout this chapter.

3.2.1   Fingerprints

Fingerprints, perhaps more than any other biological feature, are associated with human identification. Specifically, they are historically associated with the identification of criminals. Thus fingerprinting has a social stigma attached to it and this has been an issue in the adoption of fingerprinting as a means of obtaining biometric identification in the general population: people feel that they are being criminalised. Within lay consciousness, fingerprints have been thought of as infallible as a means of human identification – a view enshrined in the popular media (Cole, 2000). Fingerprints have been part of human consciousness for thousands of years, with evidence of their reproduction ranging from rock carvings in Neolithic France and Nova Scotia to use as a unique trademark or seal in ancient Babylon and China (Ashbaugh, 1999). Fingerprints are essentially a form of friction ridge (others include the ridges on the toes and lips), and their potential to distinguish one person from another has been known for hundreds of years. Surprisingly, given fingerprints' ubiquitous nature in crime scene and forensic investigations, scientists have conducted little in-depth research into them.

Fingerprints are not genetically predetermined, but, as with most aspects of the human body, their form arises from a combination of genetics and foetal and maternal interactions with their respective environments. They form during the development of the foetus in the womb. The epidermis will develop from around four weeks following conception, while the dermal layer will appear beginning at eleven weeks (Gawkrodger, 2002). The hand will begin to develop at around five to six weeks and then, by six to seven weeks, the fingers will start to appear (Champod et al., 2004). By six weeks of development, the foetus will respond to physical stimulation of its skin surface by moving away from the source (Montagu, 1978). Fingerprint ridges will be ultimately developed by twenty weeks in utero (Gawkrodger, 2002; Champod et al., 2004).

The friction ridges themselves develop from the volar surfaces (tips) of the digits. These regions appear by around seven weeks of foetal development and occur first between the digits and at the edges of the palms and then will subsequently develop on the distal ends of the fingers by seven to eight weeks (Schaumann & Alter, 1976; Babler, 1987; Champod et al., 2004). These swellings lose their prominence by sixteen weeks as the growth of the hand overtakes the pads (Champod et al., 2004). The volar surfaces of the fingers of primates exhibit regularly spaced papillary ridges and grooves, and these are associated with the superficial stratum corneum of the epidermis (Lemelin, 2000), although they begin to develop from proliferating cells at the basal layer (Babler, 1987; Champod et al., 2004). These ridges and grooves are a feature of the epidermis–dermis junction (Lemelin, 2000) and as they mature they extend deeper into the dermal layer (Champod et al., 2004). Therefore, these primary ridges and furrows on the epidermal surface correspond with those of the underlying epidermis–dermis junction (Schaumann & Alter, 1976). When they begin to appear, these ridges develop in three key regions: the tip of the finger, the core of the finger and the interphalangeal joint (Champod et al., 2004). Once the primary ridges and grooves are established, further ridges and furrows form. These are referred to as secondary ridges and are also present at the epidermal–dermal junction. Occasionally smaller, narrower and fragmented incipient ridges can be found between these normal ridges (Ashbaugh, 1999). Together, these primary and secondary ridges and grooves form into distinctive patterns and are subsequently referred to as 'fingerprints' (Lemelin, 2000). Despite their continued use in routine forensic investigation, the actual physiological processes that result in our fingerprints are still debated (Kücken & Newell, 2005). Nonetheless the formation of fingerprints is the same in everyone, although the actual prints are unique, even in identical twins. This is because the uterine environment acts upon the fingers during development.

Ultimately what provides a fingerprint with its identification potential is the presence of general patterns and minutiae (see Figure 3.1). These can be categorised into three levels and are detailed in a number of other publications (e.g. Jain et al., 2002; Champod et al., 2004). Level 1 features the overall trend patterns formed by the primary ridges, including simple and tented arches, right and left loops and whorls (Figure 3.1). These general patterns are thought to be indirectly inherited in that they are influenced by a number of factors, including size and shape of the volar pads, timing between regression of the volar pads at around sixteen weeks and the formation of the primary ridges, relative speed of the three development fronts and bone morphology (Jain et al., 2002; Champod et al., 2004). Level 2 features include major ridge deviations such as ridge endings, bifurcations and dots. Pathologies, scars and folding creases are also deemed Level 2, although as Ashbaugh illustrates, such pathologies can be temporary, such as a wart, which can appear for a prolonged period of time but then regress leaving the ridges the same as before (1999). Level 3 features include intrinsic ridge formations, alignment and shape of ridge units and pore position and shape. Fingerprints contain between 75 and 175 useful ridge characteristics (Epstein, 2002); key to the identification process is the fact that these minutiae are not evenly distributed (Champod et al., 2004). One can think of the progression from Level 1 to Level 3 as 'zooming in' on the fingerprint, or moving from a more macroscopic to microscopic examination.

It is important to note here that, in this field, a direct association is made between 'identification' and 'individualisation' (Champod et al., 2004), even though from a conceptual perspective, these two are not the same. Techniques have developed since their initial usage in crime science and today the key method is known as the ACE-V approach (Analysis, Comparison, Evaluation and Verification). Several workers have summarised the main stages of the

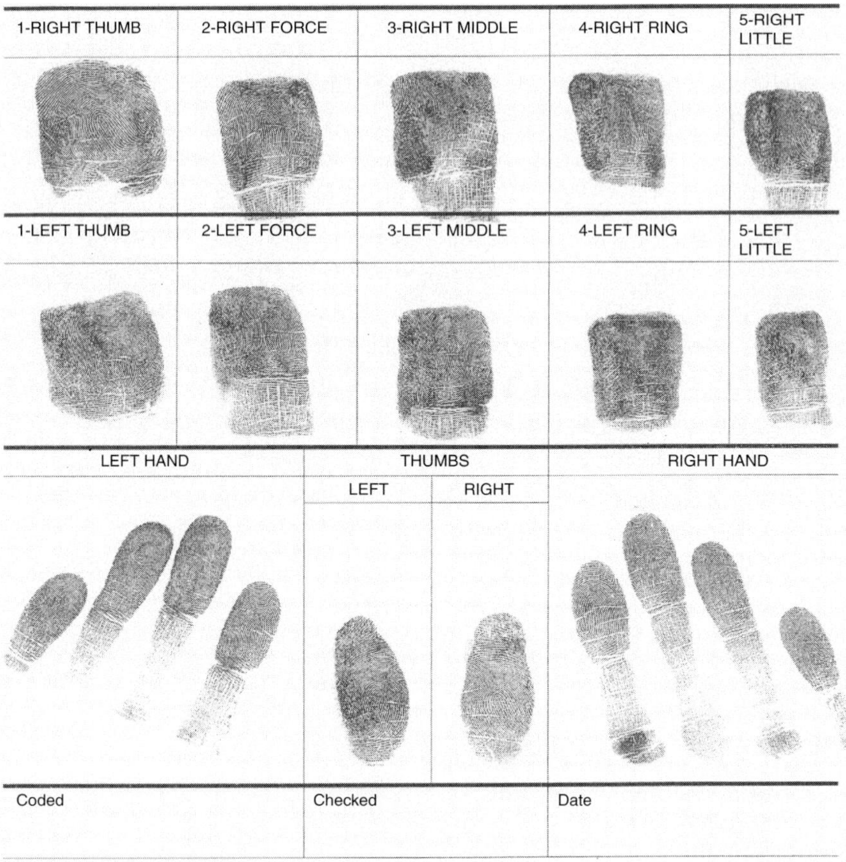

1-RIGHT THUMB	2-RIGHT FORCE	3-RIGHT MIDDLE	4-RIGHT RING	5-RIGHT LITTLE
1-LEFT THUMB	2-LEFT FORCE	3-LEFT MIDDLE	4-LEFT RING	5-LEFT LITTLE

LEFT HAND	THUMBS		RIGHT HAND
	LEFT	RIGHT	
Coded	Checked		Date

Figure 3.1 Examples of friction ridge patterns from fingers. Courtesy of Ian Parker.

ACE-V approach (e.g. Champod et al., 2004), and these are highlighted in Table 3.1. The key limiting factor influencing all of the questions in the Analysis column of Table 3.1 concerns the way in which the print was deposited: how long contact was made for, whether the mark was left in a fluid or on a porous surface, whether distortion of the mark had occurred, the impact of the development technique used and so on. Although the Comparison and Evaluation stages (Table 3.1) focus on detecting similarities, in reality, the differences are more significant as a single variation is enough to note an exclusion (Champod et al., 2004). In addition, it must be borne in mind that although our finger-prints seem stable on our hands, a number of intrinsic (biological) and extrinsic (environmental) factors will influence the quality and detail left by a contact (Cartmill, 1979; Fieldhouse, 2011). Interestingly, even extremely dehydrated and mummified remains can produce excellent prints if they are suitably

Table 3.1 *Key questions derived from the ACE-V methodology*

Analysis	Comparison	Evaluation	Verification
Is the mark a friction ridge impression?	Is the print (note it is no longer a 'mark') of a sufficient quality for comparison, and if it is not, is it possible to acquire additional prints?	Is there agreement with the Level 3 characteristics?	Does another practitioner agree with the conclusion?
What are the medium conditions surrounding the fingerprint?			
Is the mark of a sufficient quality to allow for a comparison to be attempted?	Is there agreement with the Level 1 characteristics?		
	Is there agreement with the Level 2 characteristics?		

Source: Adapted from Champod et al., 2004.

treated, for example, by excising the epidermal, dermal and fat layers of the tip of the finger and gently rehydrating them (Fields & Molina, 2008). The ACE-V philosophy may be universal, but the specific approaches to achieving this are not. Two key methodological camps work in the forensic field today. The first argues that identification can only be possible if an agreed number of concordant minutiae are located between the mark and the known print. The second argues that a more holistic approach is required and that both qualitative and quantitative assessments are necessary. Cole (2000) and Epstein (2002) highlight the lack of a scientific basis for the numerical minimums of concordant points. As with other areas of human identification, the use of probabilities has been suggested for fingerprint analysis. There is reluctance to adopt this approach, however, as some scientists and investigators working in the field argue that a fingerprint either matches or does not and that there is no in between (Champod et al., 2004).

As mentioned earlier, the uniqueness of fingerprints extends to identical (monozygotic) twins. Fingerprints do display genetic similarities, and it has been argued that these are strong enough to aid in establishing paternity (Schaumann & Alter, 1976). In reality, however, these genetic influences affect fingerprint patterns because genes determine the location of the volar pads in addition to the overall size and shape of the digit and the bony skeleton beneath (Babler, 1987). Work has also shown that the width of the volar pad is the most influential on pattern, rather than the height (Babler, 1987). The specific changing micro-environment surrounding each foetus is, however, unique enough to result in differing fingerprints. Nonetheless, there is likely to be a relatively high level of minutiae similarity, as these will naturally derive from pattern similarities seeing as the minutiae are associated with ridge location (Jain et al., 2002).

Forensic human identification using fingerprints is one of the most established and respected forensic tools, to the extent that it has successfully passed the infamous Daubert test – a means of determining the validity of forensic sciences. Fingerprints' first use in the judicial context was arguably during Hammurabi's rule of Egypt between 1792 and 1750 BC (Ashbaugh, 1999), while the state of Illinois was the first to admit them as evidence in the United States in 1911 (Epstein, 2002). Despite this general acceptance, an increasing number of forensic and legal practitioners are beginning to question the appropriateness and validity of this form of evidence. In the United Kingdom, this came to a head when Shirley McKie, a detective constable in Scotland, was charged with perjury after her fingerprint was found contaminating a crime scene – yet after fierce debate within the fingerprint world for a number of years, she was exonerated (Broeders, 2006). These concerns stem from a number of issues, summarised by Cole (2000) and then Epstein (2002): (a) that no one has really demonstrated that everybody's fingerprints are unique; (b) that no measures for the reliability of fingerprint evidence are in place, and no error rates are in existence – although Christensen and Crowder have highlighted the rather rash claims of some practitioners that fingerprint examiners have an error rate of zero (2009); Grivas and Komar question whether error rates can even be established for such unique features (2008); (c) that latent prints averaging one-fifth of the size of a complete fingerprint are being examined, and again, no work has been undertaken to assess the degree of uniqueness in the population of partial prints; (d) that the ACE-V methodology does not provide experts with information regarding the evaluation or verification aspects of examination; and (e) that a peer-reviewed body of literature in scientific publications does not exist (although it is countered that fingerprint evidence is peer reviewed through the courts, this process rarely addresses the underlying concerns regarding error or uniqueness of prints, neither do the majority of academic publications on fingerprint evidence). Thus Epstein argues strongly that fingerprint evidence fails to meet the 'scientifically valid' demands set out by the Daubert hearing regarding the admissibility of forensic evidence and expert testimony (2002). Other authors are less scathing though, with Christensen and Crowder noting that subsequent rulings acknowledge that not all of the Daubert criteria will apply in all cases (2009).

In addition to the forensic applications, the 'uniqueness' of fingerprints is being incorporated into more everyday contexts. Naturally, fingerprints are still used to determine or verify identity, but the scenarios focus more on access to mobile phones, laptop computers, cars and property. Even USB memory sticks can incorporate fingerprint-access technology. The use of fingerprints is also discussed in relation to accessing state benefits (and is already required to access welfare in the United States), with the implications that this naturally has (Kruger et al., 2008). Regardless of the sheer scale of public, judicial and technological acceptance of the importance of fingerprints as a method of human identification, their uniqueness is still not entirely recognised.

In addition to the fingerprints themselves, the ratio of relative finger lengths when handprints are reproduced has been used as a means of sex and age identification in forensic and archaeological cases. An example of the latter includes the work by Van Gelder and Sharpe, which has sought to establish the identity of prehistoric artists through the measurement of the preserved handprints (2009). Some researchers have argued that relative finger lengths are sexually dimorphic and that this dimorphism is apparent from early childhood. Using this methodology, Van Gelder and Sharpe have demonstrated that a substantial proportion of prehistoric artists were female (2009). Following from this, Jessica Cooney (pers comm) has recently shown that some prehistoric art was the work of young children. Such studies successfully challenge the andro- and adult-centric views of the past that have previously characterised much archaeological discourse.

Beyond personal identification and verification, fingerprints and palm creases can be used to suggest potential medical disorders in individuals. This is because a number of genetic disorders manifest at a similar time to the embryological development of fingerprints, thus interrupting their development. Schaumann and Alter describe the typical dermatoglyphic effects of these illnesses, and they include ridge aplasia, ridge hypoplasia, ridge dissociation and ridges-off-the-end (not restricted to the pads of the fingers and running over the borders) (1976). Genetic illnesses that can affect fingerprint morphology include Down's syndrome (i.e. by creating vertically oriented and L-shaped loops and reducing friction ridge counts) and sex chromosome defects (i.e. increased number of sex chromosomes reduces friction ridge counts by influencing finger fat pads). The influence of Down's syndrome is of interest considering that despite improved methods of detection, there is currently an increase in the number of children born with Down's syndrome in the United Kingdom, arguably due to the lessening of social stigma, improved quality of life outcomes for those with the condition and the occurrence of pregnancy at older ages (see Tringham et al., 2011 for greater discussion on Down's syndrome screening in the United Kingdom). Even non-genetic-based illnesses, particularly in the mother during pregnancy, can affect fingerprint development. Examples include rubella, leukaemia and celiac disease – although some of the fingerprint effects of these environmental illnesses can be reversed, such as with an alteration to a gluten-free diet of those suffering from celiac disease (Schaumann & Alter, 1976). As has already been stated, fingerprints are not genetically predetermined but are influenced by a combination of genetics and foetal and maternal interactions with the surrounding environment.

Friction ridge patterns on the fingers and hands have also been used in an attempt to move beyond individualisation in order to explore human identity more broadly. The functional differences between primates and other mammals have resulted in morphologically different extremities and these differences extend to the friction ridges. Specifically, the morphological differences of the volar skin have been linked to the action of grasping and locomotion with the volar pads of some species being discrete and elevated rather than blending into the digit as they do in humans (Cartmill, 1979; Lemelin, 2000). Furthermore, the climbing medium itself may have a part to play so that ridges develop to a greater extent if the branches are smaller in diameter (Hamrick, 1998). Lemelin has argued for the use of fingerprints in mapping human and primate evolutionary pathways in a way that opposes current biomolecular approaches (2000). The key observation used is that those mammals that rely on active friction grips for locomotion (e.g. primates and arboreal mammals) have much more defined papillary ridges and grooves with a more undulating epidermis–dermis junction than those that, for example, fly. Similar arguments were made 20 years previously, where the coalescing of the volar pads into the digits was seen as a sign of 'phylogenetically higher development' (Cartmill, 1979: 508). Study of fossil samples has allowed the production of new phylogenetic trees for primates and humans, allowing for a new interpretation of humanness in general (Lemelin, 2000).

3.2.2   Palmprints

Ashbaugh highlights the potential of the flexion creases on the palm of the hand to individualise people (1999). These creases appear to mark regions of firmer skin attachment to deeper regions of the hand, and they emerge on the foetus earlier than friction ridges. They are, therefore, stable morphological features, although it should be noted that some smaller creases may be affected by age-related dehydration and the accumulation of fat. Other authors (Chen et al., 2010) add that their usefulness also lies in their ubiquity, permanence and ease of collection. Essentially, the palmar area of the upper limb is divided into topographical areas based on the location of the volar pads (Ashbaugh, 1999). The creases are termed Major (the three main lines that run perpendicular to the axis of the limb), Minor (the three that run from the wrist to the end of the fifth to third metacarpals, one that runs along the heads of the fifth to third metacarpals, the collection at the base of the fifth digit and one that runs parallel to the radial edge of the palm) and Secondary (the remaining smaller, shallower creases across the palm). Genetic control of the flexion creases is similar to that working on fingerprints, in that the position of the volar pads and underlying skeletal structure influences positioning. Monozygotic twin studies have demonstrated that although Major creases may follow the same orientation, there is enough divergence within the Minor creases to allow for determination of uniqueness (Ashbaugh, 1999).

Those interested in determining the probability of certain genetic illnesses in foetuses have adopted the examination of the morphological features of the palm. This is because these flexion creases develop concurrently with the manifestation of these diseases (Ashbaugh, 1999).

In the same way that partial prints can be exploited in fingerprint analysis, partial palmprints can be exploited for identification purposes; indeed, some researchers have argued that the larger size and greater robustness of this region gives palmprints distinct advantages over fingerprints (Chen et al., 2010). In their preliminary work, Chaudhry and Pant highlight the importance of the palm as a support during writing (2004). They note that the partial print of the radial side can indicate not only if the author is right or left handed (convex versus concave print), but also his or her approximate age (based on size) and his or her identity if creases or friction ridges are present. More recent research has shown that the use of low-resolution imaging produced identification accuracies of 99.90 per cent for the palm (Chen et al., 2010).

The influence of palmprints on social identity comes from two key directions. The first is when, as mentioned earlier, the creases of the palm are used for predicting the likelihood of being affected by a given illness. Similar issues surround the use of genetic testing for diagnosing various disorders (see Chapter 6 for far greater discussion). Beyond this, palms have been used to comment on the future lifeway of a person through the application of palmistry. This is a practice with a long history, based on the notion that every individual has unique detail in his or her palm creases through which information can be read. Despite its continued use, it was dismissed in the fourteenth century (along with some other superstitions) as 'not sciences properly speaking' (Carey, 2010).

3.2.3   Footprints

The foot is designed to absorb the weight of the body and to counter the actions of gravity and shock resulting from locomotion (Kennedy, 1996). Due to the complex anatomical nature of the foot, it has been argued repeatedly that it is highly unlikely that two individuals share the same overall structure, including identical twins (Bodziak, 2000; Kennedy et al., 2005; Krishan, 2007). In his influential text, Bodziak details the sources of foot-related data and cites the army, the footwear industry, sports and medical laboratories as key locations of research (2000). Nonetheless, it is clear that far less research has been conducted into the issues of footprint identification compared to the hand and fingers.

The curvature of the plantar surface of the lower limb is thought to be unique to humans, and the extent (including the lack) of the arch is a result of loose ligaments and joints (Echarri & Forriol, 2003). The arch is designed to help support the weight of the body, but the seven tarsal bones absorb the majority of the pressures, with the metatarsals assisting with this task and providing needed leverage for walking (Robbins, 1985). The lack of an arch, or flat-footedness, has provoked much interest over the years because it has interesting cultural and economic ramifications. In India, for example, the condition prevents recruitment into the military and police (Krishan, 2007). The arch was initially argued to be the preserve of 'civilised people' (James, 1939), but more recent ethnographic research suggests that barefoot locomotion actually reduces the likelihood of flat-footedness (Tuttle et al., 1990; Echarri & Forriol, 2003). Examination of eighteen hundred fifty-one Congolese children by Echarri and Forriol suggested that most feet are flat at the ages of three and four, but that increasing age brings a reduction in the amount of flat-footedness (2003). This is supported by Kulthanan et al. (2004). Sexual dimorphism is present, with boys being more susceptible than girls, while the arches in girls were generally higher in the Congolese group. Flat-footedness is also thought to exhibit bilateral variation (Krishan, 2007), however it has yet to be related to the fact that the sizes of the bones and feet also show bilateral variation depending upon side dominance (Robbins, 1985). A barefoot existence also results in a widening of the space between the first and second pedal digits and a broader foot in general (James, 1939; Tuttle et al., 1990). These morphological differences, which are essentially the result of cultural practice, mean that the foot functions in slightly different ways during locomotion, with regularly shod feet placing maximum walking strain onto the first metatarsal rather than spreading the load across the set, making little use of the toes during push-off and placing strain along the lateral border of the foot rather than across the arch (James, 1939).

Interestingly, workers have failed to correlate the nature of the arch to activity, such as jumping, running or weight bearing (Kulthanan et al., 2004), although metric measures of the foot indicate that obesity correlates with lower arches and flat-footedness (Dowling et al., 2001; Echarri & Forriol, 2003). Robbins argues that body size is related to the pressures we place on our feet, although this is not quantified and may be an assertion of common sense (1985). Dowling et al. record that, although obesity places greater pressure on the foot, the amount of the foot in contact with the ground is greater in obese people, therefore keeping overall pressure per unit similar to non-obese people by distributing it over a larger surface area (2001). This is in contrast to non-obese people who carry heavy objects. Their research suggests that long-term weight bearing will cause structural alterations to the foot, while short-term weight bearing will not increase pressure per unit in the latter category. Nevertheless, it is unclear from their data whether the structural changes resulting in an increase in plantar surface area were the result of changes to the longitudinal arch, differences in foot length/width ratios or the presence of a fat pad in the foot (Dowling et al., 2001).

Arguably the most well-known trail of human footprints is that from Laetoli in Tanzania. Discovered in 1978, they represent the impressions of approximately sixty-nine bipedal footprints extending over a distance of 27.5 m. They include the impressions of three individuals, one small and the other two taller (Hay & Leakey, 1982; Tuttle et al., 1990) and, although the impressions of the larger individuals overlap, they show a rounded heel, elevated arch and forward-pointing large toe akin to modern humans; all the tracks suggest a normal bipedal gait (Hay & Leakey, 1982). Comparative work by Tuttle et al. on indigenous Peruvian populations suggests that the three individuals recorded at Laetoli had similar pedal features (1990). Radio-isotopic dating places the footprints to between 3.8 and 3.5 million years ago (Hay & Leakey, 1982). The main scientific interest at Laetoli rests on the bipedal footprints – although it should be noted that most work has been undertaken on casts rather than the actual specimens – they actually form part of sixteen sites of footprints in this region which include animal tracks and, perhaps most remarkable, depressions resulting from raindrops. Hay and Leakey explain that the exceptional preservation in this region is due to geological factors, including the nature of the soil, the degree of hydration and the presence of a covering layer of volcanic ash (1982). Stratigraphic analysis suggests that the human and animal tracks made here coincide with wet season migration.

There are interesting parallels between these ancient footprint trails and more modern ones. Like Laetoli, the preservation of the footprints of Neil Armstrong on the moon is significant in terms of the identity of humans and humankind, their development and endeavours. Although less well publicised, perhaps due to their apparent invisibility, similar cultural heritage issues face the impressions and other artefacts resting on the moon (Rogers, 2004; Spennemann, 2004, 2006). But as Spennemann states:

> There are defining events in everyone's personal life, and there are
> defining events in a nation's history that occur during a lifetime … But

then there are those events that define, consciously or subconsciously, a large part of humanity, irrespective of generations, ethnicity and creed. (2006: 356)

From an identification perspective, footprints can be recovered from a variety of forensic situations. Examples presented in the literature include blood, mud, inside shoes and casts. Socked impressions can also be used (although the clarity of the impression can be reduced; Bodziak, 2000). Casts can also be recovered from impressions in the snow, although care must be taken not to use casting agents that produce a thermal reaction during hardening. Cultural, religious and climatic considerations are important here too, as in some countries, such as India, the importance of footprint analysis is high due to the high levels of unshod locomotion (Krishan, 2008). As with all footprints, it is important to remember that examination focusses on the negative image of the plantar surface of the foot (Robbins, 1985; Krishan, 2007). Methods of identification or profile construction focus on three areas: friction ridges on the plantar surface, the dimensions of the foot and unique morphological and pathological features on the foot. The first will not be examined in any detail here, as the principles are the same as with fingers, other than to say that, like the hands and fingers, the feet and toes develop volar pads and ridge systems, albeit developmentally later than in the upper limbs (Champod et al., 2004). Plantar creases can also be used in a similar way to the palmar creases of the hand (Schaumann & Alter, 1976). Dimensions of the feet can be recovered from both within the shoe and as a mark deposited on a surface. Within the shoe, impressions of the foot are left behind in sweat stains and as depressions (Robbins, 1985; Kennedy, 1996; Bodziak, 2000); it can take as little as forty hours of wear for an impression to become visible and 150 hours for the impression to become of a quality suitable enough for forensic comparison. Much work has been conducted investigating the potential of footprint dimensions to aid identification. Robbins discusses the use of the length and width of various anatomical features, such as maximum length, heel-to-ball, heel-to-arch, toe length, arch width and so on, in addition to the angle created between similar features (1985). Kennedy's early work used a database of four thousand impressions from two thousand individuals (1996). He concluded that it only took three to five measurements to exclude all other footprints from the matching print, and that even with an error margin of 5 mm, twelve to fifteen measurements sufficed to determine uniqueness. The notion that small numbers of metric measurements are sufficient to distinguish impressions is supported elsewhere (Bodziak, 2000). Later work exploiting an automated system and a sample of five thousand seven hundred fifty-five individuals suggested that the probability of chance matches within a general population was 7.88×10^{10}, or 1 in 1.27 billion (Kennedy et al., 2005). Many morphological features of the footprint can be utilised, including the shape, size, position and angle of the toes (particularly if any are missing), the presence of toe stems (rarely seen in a footprint), the nature of the ball area, the extent of the arch and the nature of the heel (Mathieson et al., 1999; Bodziak, 2000; Urry & Wearing, 2005; Krishan, 2007). Unique pathological features can include ulcers, corns, cuts and fungal infections (James, 1939; Krishan, 2007).

Anthropologists have also been keen to associate footprint dimensions to bodily proportions. Most simplistically, this has been done with rough approximations, such as in Hay and Leakey, who suggest that footprint length is 15 per cent of height (1982), and Kulthanan et al., who note a similar 7:1 footprint to height ratio (2004). More rigorous statistical approaches have also been adopted. Correlations have been devised for footprint dimensions with weight and with stature, although the latter have more accuracy. Robbins' study revealed that correlations with stature tend to be stronger when the lengths of pedal components are used rather than widths (1985). She also noted that, as discussed previously, the toes can be problematic. Equations for predicting weight tend to suffer due to the fact that body dimensions are not as strongly influenced by weight as height, that weight is not as stable as height and because increased weight causes soft tissue deformation to the foot and thus the print it leaves (see Krishan, 2008 for quantification of this). Much population variation exists within these prediction equations, and care must be taken to ensure that a suitable equation is used for the footprint under examination.

3.2.4   *Earprints*

For many years, earprints have held a great allure for forensic investigators. Since the early work of Alphonse Bertillon in the nineteenth century, forensic investigators have believed that the folds of the ear must be unique to each individual, and therefore successful recovery from a given context would allow them to link a scene to a person. This assumption regarding the uniqueness of earprints is similar to that regarding fingerprints, in that it cannot be fully tested or proven (Burge & Burger, 1999). For Bertillon, a student of pioneering forensic scientist Locard, the analysis of the ear for individualisation formed part of his System of Identification (Signaletic Instructions). It featured in all three levels: Anthropometrical, Descriptive and Peculiar Marks (Chapman, 1993). Despite initial and resultant interest, little research has been undertaken in this field. In an attempt to answer the question of the significance of earprints in human identification, the FearID project was undertaken, funded by the European Union. Involving researchers in the Netherlands, Italy and the United Kingdom, the project aimed to assess the uniqueness of

earprints and to determine their scope of individual identification and verification. Initial output of this collaboration (Meijerman et al., 2004) was to
highlight the embryonic nature of this field of study, noting that fundamental questions as to the degree of similarity between ears, the impact of pressure on print quality and so forth had not been fully addressed. At the time,
it was still not known whether inter-individual variation in print was greater
than intra-individual variation. Indeed part of the problem was that the flexible
nature of the auricle (ear) meant that different anatomical features responded
differently and fairly unpredictably due to increases in pressure, while the
three-dimensional nature of the structure meant that some features did not
make contact with the surface on all occasions. Other problems investigators
and researchers have addressed include the occasions when hair and so forth
covers the ear (although this could be resolved with the application of thermal
imaging; Burge & Burger, 1999), the fact that the auricle was not actually stable
during the life course and that it changed dimensionally with age, although
this was deemed to occur over such a long period of time as to be of little forensic significance (Meijerman et al., 2004). Indeed most growth occurs between
birth and four months and then after 70 years (Burge & Burger, 1999), arguably
outside the general age range of individuals of forensic interest. Between these
ages, growth is generally proportional, which would not restrict morphological
methods of identification (Burge & Burger, 1999). Nonetheless, both metric and
morphological methods for analysis are present in the literature, although they
all tend to focus on the helix, antihelix, tragus and antitragus structures.

This in itself has proven problematic, as subsequent studies have compounded the difficulties of using earprints for forensic investigation by highlighting the high degree of inter-observer differences in merely annotating
anatomical regions and morphological features from the latent prints (Alberink
et al., 2006). As the authors argue, the correct annotation of the earprint of
interest is the starting point for all subsequent analyses. Although only a small
number of operators were used for comparison (three), the paper points to a
worrying feature of earprint examination.

The lack of significant scientific research and debate has not stopped the use
of earprints in the courts. The key case in the United Kingdom is that of *R v.
Dallagher*. Initially convicted based on the presence of his earprint at the scene
of a murder, Dallagher's conviction was overturned on appeal after analysis of
the DNA within the print proved it belonged to someone else (Broeders, 2006).
The fallout was, as one would expect, severe, and has led to an abandoning of
earprint work. Subsequent work by Graham et al. has shown, however, that
it is possible to recover DNA from an earprint that originates from a different
source to the print itself (2008). Potential causes of this include shared use of

a telephone, for example. Arguably this raises further questions regarding the Dallagher case, as a consequence of which the use of earprints in the UK court system seems to be in turmoil (Broeders, 2006).

3.3   **Skin and identity**

> Skin differentiates but does not isolate. Your singular existence unfolds within it, but skin does not hold the universe at bay. Instead it marks the seam that joins your existence to everything else. (Jones, 2007)

Our skin is the interface between ourselves and the world. Much has been written about the liminal position of skin, the internal–external border zone it occupies the role it plays in social identity and identification and the way in which individuals adorn and manipulate the skin as a means of indicating group or individual identity, cohesion and difference. Benthien has argued that in modern times the skin has represented an increasingly rigid boundary between self and other, and this despite it being frequently penetrated by medical science and what lies beneath being better 'known' than ever before (2002). The skin is impregnated with our personal histories (Prosser, 2001), and becomes constituted by our social relations, inscribed by our lives and interactions; as a highly visible organ, it, in turn, moulds these relationships.

3.3.1   *Age*

The skin 'materialises the passing of time' (Ahmed & Stacey 2001: 2) or, as Connor puts it, the 'skin is written by time' (2001: 48). The temporal component to the skin is of considerable concern to many keen to thwart and deny the physical effects of ageing amongst Western culture's youth fetishists. Calleja-Agius et al. describe seven different classifications and causes of physical ageing (2007). These are chronological, genetic, photo ageing (the influence of UV radiation), behavioural (smoking, alcohol abuse, etc), catabolic (diseases such as infections and cancers), endocrine (changes in hormone levels) and gravitational. There is a strong relationship between the mechanical properties of skin and the age of the individual it covers. Despite the skin's prominence, scientists have conducted little research into its mechanical properties. The reasons for this clearly focus on the difficulties of studying a living tissue. Much current work focusses on the application of finite element analysis (e.g. Bischoff et al., 2000; Delalleau et al., 2008), while in vivo studies tend to deploy ultrasound modalities (e.g. Escoffier et al., 1989; Diridollou et al., 1998; Petrofsky et al., 2008). Other attempts to assess in vivo skin properties have been undertaken, but they have not been as successful (such as Falanga and Bucalo's application of the durometer to test skin hardness across the body; it failed in anatomical regions with thin skin thickness [1993]).

In addition to changes in mechanical strength, ageing skin becomes more rigid from the age of 60 years onwards (Escoffier et al., 1989). Flattening of the basal layer of cells can also be seen, which consequently makes this region more distinct histologically (Oriá et al., 2003). Changes to the mechanical properties in the skin tend to result from age-related changes to collagen structures and the dermis, such as fragmentation of the elastic fibres (Oriá et al., 2003; Petrofsky et al., 2008) or increased cross-linkages within the collagen (Bischoff et al., 2000). A review of the field also suggests that collagen within the skin fragments over time, although the tight intermolecular cross-links it associates with are highly resistant to complete breakdown, and cannot be repaired or reincorporated into the collagen fibrils (Fisher et al., 2008). This in turn compromises structural integrity by reducing the degree of mechanical tension possible on the fibroblasts, which impairs fibroblast functioning and reduces collagen production, as they both depend on tension applied to the fibroblasts for efficient functioning. This then becomes a feedback loop from which there is no recovery. This loss of tension will be exhibited through the sagging of skin; particularly noticeable in those more vertical parts of the body such as the orbits of the face (Sforza et al., 2009). This is referred to as *ptosis* (Clarkson & Schaefer, 2007). There is also thought to be an important hormonal control on this process. Evidence for this comes from the fact that following menopause, 30 per cent of skin collagen is lost in the next 5 years and that the application of oestrogen can prevent this decline (Calleja-Agius et al., 2007). Contrary to this, however, Castelo-Branco et al. find a stronger correlation with chronological age than with time since menopause, but this could be a sampling issue (1994). This theory concerning the transformation of the skin via changes in collagen is argued to be preferable to the well-publicised free radical theory, which states that the key force of ageing is the presence of reactive forms of molecular oxygen which are produced via aerobic energy metabolism and then oxidise, impairing the function of the cellular components (Fisher et al., 2008).

Damage to the DNA in mitochondrial cells has been highlighted as a potential cause of age-related changes to the skin, primarily because such damage will affect the efficacy of mitochondria to perform respiration (Khrapko & Vijg, 2009). Mitochondria are found in large numbers in muscle cells, and therefore damage to these organelles might preferentially affect these tissues – and indeed, this seems to be the case since mild increases in mitochondrial DNA in muscle tissues has been seen to present muscle-wasting phenotypes. In fact, muscle atrophy can result in 40 per cent of muscle mass disappearing by the age of eighty (Khrapko & Vijg, 2009). The same authors concluded that the situation and causal relationships are far from clear. More discussion on mitochondrial DNA can be found in Chapter 6 later in this book.

In addition to changes in tissue strength, ageing also results in changes to the layers of the skin themselves. Work has shown that the stratum corneum and the dermal layer grow thinner with age (Branchet et al., 1990; Oriá et al., 2003; Petrofsky et al., 2008). It should be noted that there is not a linear relationship between age and skin thickness, rather thickness increases from birth to around 15 years, where it remains constant until around 65 years of age. This reduction in old age continues until the thickness of skin in a 90-year-old is thinner than that of a 5-year-old (Escoffier et al., 1989; Oriá et al., 2003). There is also the suggestion of some sex differentiation in the reduction of these tissues. Research with biopsies from healthy skin has shown that while reduction in dermal thickness occurs at a reasonably constant rate in men and women, it is statistically faster in men in the epidermal layer (Branchet et al., 1990). Changes in skin thickness derive from decreases in collagen, water and glycosaminoglycans (Castelo-Branco et al., 1994; Calleja-Agius et al., 2007). Changes to the epidermis–dermis junction can also be seen, with a loss of the vertical fibres that help connect the two together (Oriá et al., 2003). The papillary dermal layer in this region also seems to display a slight thickening following the initial age-related reduction (Branchet et al., 1990). Finally, the layer of subcutaneous fat also thins, although this tends to be redistributed into the abdominal region of the body. Although changes in skin thickness occur somewhat rapidly, changes to the functional ability of the skin occur gradually – and from a much earlier age. Such changes include the elasticity of the skin and the time taken for skin to relax after strong deformation (Escoffier et al., 1989). Furthermore, with a thinning of the skin comes a reduction in functional vascularity (Petrofsky et al., 2008). Actual blood flow through the skin is also influenced by age-related impairment of nitric oxide synthesis that assists dilation of the smooth muscle in the blood vessels (Holowatz et al., 2007; Petrofsky et al., 2008). Regardless of the cause, a reduction in blood supply to the skin will impair function and repair, while increasing the risks associated with exposure to extreme temperatures.

The tactile sensitivity of skin also decreases with age along with auditory and visual acuity. This change has been found in both males and females and is possibly related to the nervous system rather than the changing mechanical properties of the skin (Woodward, 1993). Given the importance of touch in our everyday interactions, this loss can have important repercussions for our psychological and physical well-being. Tactile insensitivity can serve to compound the social marginalisation felt by many elderly people. If we are numb to the touch of the world and our other senses are similarly clouded, then, instead of vital and vibrant means of stimulus and information transmission, our bodies become cocoons.

Throughout our life course, our skin can accumulate scars of physical trauma and pathologies associated with the passage of time, rather than specific events per se. Some of these are the result of continuous exposure of the skin to the environment over the life span of an individual. Such extrinsic forces include ultraviolet radiation and levels of nutrition (Oriá et al., 2003; Calleja-Agius et al., 2007). Other age-related traumas and pathologies, for example pressure sores and ulcers, tend to be the product of reduced mobility and are a common feature of older age. Oftentimes they are also correlated with a lack of proper care or with neglect and abuse. The sacrum and pelvic region have been shown to exhibit more pressure sores than other regions of the body. Work by Keelaghan and colleagues has shown that those older individuals who live in nursing homes are more likely to clinically present pressure ulcers when entering hospital than those who enter from family homes (2008). The caveat here is that those who live in nursing homes are at a greater risk of suffering pressure ulcers as they tend to be placed in such homes precisely because they are less mobile and cognitively impaired (Keelaghan et al., 2008). Nonetheless, these age-related changes to the skin result in a reduced capacity of the skin to heal itself following injury.

Perhaps the key factor concerning age-related changes to the human skin is the fact that these changes are so visible, and as such, are significant in terms of self-identity. In Western society, we have fetishised the smooth, elastic skin of childhood to the extent that the practice of surgically lifting and stretching the skin of the face in an attempt to mimic this smoothness is becoming increasingly common (Benthien, 2002). Not only have age-related physical changes been linked to negative impressions of social potency and intellectual competency but also to positive impressions of wisdom and benevolence (Hummert, 1994; Montepare, 2006; Hebl et al., 2008). Research has shown that individuals become increasingly concerned with their external physical appearance as they age, but simultaneously view the competence of their bodies with increasing favour (Montepare, 2006). The appearance of wrinkles, the greying of hair and the sagging of skin in particular, are linked with feelings of decreasing self-esteem (Hummert, 1994; Calleja-Agius et al., 2007). Perhaps not surprising, clear sex differences can be seen here, with women demonstrating greater body consciousness with increasing age than men (Montepare, 2006). Interestingly, there does not seem to be a simple relationship between actual chronological age and subjective age. Younger individuals tend to associate themselves with an older subjective age identity

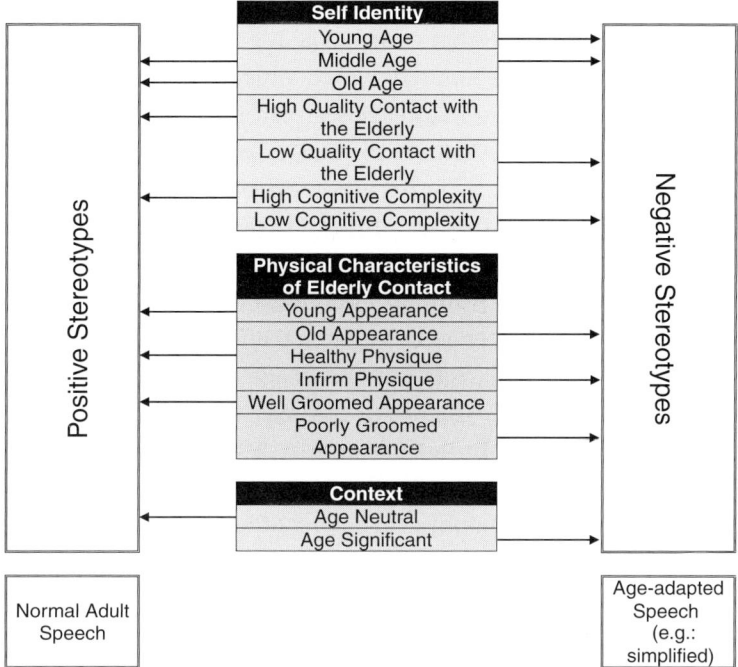

Figure 3.2 The application of positive and negative stereotypes in interactions with older individuals (adapted from Hummert, 1994).

while older individuals associate themselves with a younger subjective age identity. This latter relationship may result from comparisons by individuals with various old age-related identity stereotypes (Montepare, 2006). Perceptions of age from others can also influence behaviour and communicative interaction. Research by Hummert has shown, however, that rather than treat the elderly as a single homogenous group, stereotypes vary across three categories: young-old, middle-old and old-old (1994). Test subjects associated more positive stereotypes (such as being generous, useful, loving and liberal) with the young-old category. They associated negative stereotypes (such as impairment, inflexibility, recluse and curmudgeonly) with the old-old age group. The middle-old age group received a roughly even balance of the positive and negative. This is also supported by the work of Hebl et al. (2008). Figure 3.2 presents an adaptation of Hummert's model for predicting the impact of stereotypes on interaction with the elderly (1994). As can be seen, self-identity is as important as the perception of the other converser in determining the likelihood of applying positive or negative stereotypes. Again, sex differences are present, with females seeing the greatest shift from positive to negative stereotypes with increasing age.

With the significance of age-related stereotyping and perceptions in mind, both Calleja-Agius et al. (2007) and Fisher et al. (2008) highlight the consequence of the multimillion-pound cosmetics industry which aims to reduce the visual effect of ageing on the skin. In fact, the market has grown by a third in the first 5 years of the previous decade (Calleja-Agius et al., 2007). This includes the application of oral and topical 'cosmeceuticals', prescription drugs, chemicals, lasers, abraders, injections of fillers or neurotoxins and surgery. In fact, eyelid and face lift surgery are amongst the five most popular forms of cosmetic surgery (Clarkson & Schaefer, 2007). Ultimately, however, as Hummert describes, barring invasive surgery there is very little we can do to halt the physiognomic effect of ageing on our bodies – and thus, our identities (1994). Associated with ageing and moving along the life course is the accumulation of evidence of traumas and illnesses, such as chicken pox (in the form of scars), an incautious approach to sun protection (moles and freckles) and stretch marks as a result of pregnancy or weight gain. Thus the very act of living combines with the physical ageing of our skin to ensure we become inscribed with aspects of our personal biographies (Prosser, 2001).

3.3.2   Gender

The skin metabolises sex hormones and responds to them (Giacomoni et al., 2009). These hormones influence both dermal and epidermal thickness along with immune system function. Age-related alterations in these hormone levels can change the pH of the skin and have implications for healing as well as susceptibility to autoimmune disease (Dao & Kazon, 2007). Males also have larger pores and produce more sebum than females. Men's skin is thicker than women's, though both sexes experience a decrease in skin thickness from 45 years onwards. Of key significance for the appearance of skin is the amount of subcutaneous fat. Men and women tend to store fat in different regions of the body, in the hip and thigh region in females and stomach in males. This may exacerbate bodily differences between the sexes, particularly today when excess fat is more common than not in northern countries.

The most obvious sex and gender differences between males and females with respect to the skin are the hair follicles that populate it. Facial and body hair are of considerable cultural significance with respect to gender and these norms are subjected to enormous historical variability. Both sexes must strip, groom and tame hair according to culturally accepted gendered norms. Hair and beard styles provide strong historically and culturally specific indicators of gendered identity and social status.

Studies demonstrate a correlation between sex and skin colour within most populations, with females generally exhibiting a lighter skin tone (Juzeniene et al., 2009). Scientific explanations of this phenomenon have suggested that sexual selection and infant mimicry are responsible. Recent theories, however, suggest that differences between the sexes are due to greater vitamin D requirements in females as a result of the demands of pregnancy and breast-feeding, and that this requires lighter skin. Insufficient vitamin D has long been associated with diseases such as rickets and osteomalacia, but adequate levels also reduce the incidence of a wide variety of other illnesses including heart disease, diabetes and a number of immune deficiency-related diseases (see Juzeniene et al., 2009 for a review).

Skin colour in southern India, specifically light pigmentation, is seen as one of the most important attributes for beauty (Ullrich, 2010). Jha and Adelman (2009) discuss website adverts for prospective marital partners in India where light skin pigmentation for women is emphasized. Similarly, in Japan since the 1980s, there has been a boom in skin whitening products (Ashikari, 2005). Colour stratification *within* the African American community is discussed by Hill (2002), who observed that skin colour has a significant effect on the lives of African Americans and more so on woman than men – with those of a lighter skin tone perceived as more desirable and achieving a higher status. Throughout the Western world, fair skin tone has long been associated with feminine beauty (Hill, 2002).

In Western consciousness the skin is something that women in particular need to work on, to keep it soft, blemish free and wrinkle free: flawless (Ahmed & Stacey, 2001). Hard, calloused skin is deemed masculine. In terms of body capital, the exposed skin of the face and hands must reflect our inner domination of mental and physical selves. This is not necessarily a modern phenomenon and Connor discusses how in the eighteenth century the smooth white marble of classical statues was fetishised and this pale flawlessness was deemed highly desirable (2004). This was to the extent that the deathly white pallor of consumption (tuberculosis) was seen as romantic (though deadly).

In Jane Austen's *Pride and Prejudice*, Elizabeth Bennet was castigated by Miss Bingley for her tanned appearance – which was deemed coarse and low status. This contrasts with today, when tanned skin is viewed as healthy and, until recently, an indicator of high status as it generally indicated that one could afford to holiday overseas. The 'fake tan' market continues to be a lucrative one in the Western world; females are the primary consumers of such products, though males are increasingly becoming users. Gender and skin are therefore related both biologically and socially in terms of environmental and cultural onslaughts, maintenance strategies and desired appearance.

3.3.3   *Class*

Other key research involving the skin concerns investigations into allergic reactions, and this in turn can tell us something about socio-economic status. Research has shown that the incidence of allergic reactions has increased recently, especially in children (Flöistrup et al., 2006), but that certain environments can provide a degree of protection against developing such potentially life-threatening reactions. As such, much work has been conducted on the use of skin as a tool for predicting potential allergic reactions to food and other substances (Sicherer & Leung, 2007). This occurs through the use and application of wheal skin and atopy patch testing, which places samples of the allergen on the surface of the skin and measures (in mm) the size of the reaction. One of the more interesting developments in this field has been the proposition of the hygiene hypothesis, which suggests that allergic reactions occur less frequently in individuals (particularly children) who live in environments with poorer sanitation and more restricted use of antibiotics and antipyretics (Flohr et al., 2006; Flöistrup et al., 2006; Sicherer & Leung, 2007). The same has also been seen in farming settings (Riedler et al., 2001). In essence, this suggestion contradicts the accepted assumption that a better standard of living, more money and therefore a higher socio-economic status provide for a better immune system and a healthier adult life. Large-scale research to investigate this tends to focus on the developing world, where it has been noted that allergic diseases are much less frequent than in highly urbanised settlements. Flohr et al.'s study with sixteen hundred and one schoolchildren in central Vietnam demonstrated that higher levels of socio-economic status (as exhibited by levels of parental education and ownership of goods) correlated with increased sensitisation (2006). Furthermore, decreased levels of smoking, use of gas rather than wood as a fuel, access to piped water and modern toilet facilities all resulted in higher levels of allergic sensitivity in children.

It has been proposed that higher levels of endotoxin (specifically lipopolysaccharide, which is derived from the cell walls of certain bacteria) seen in dust in the living premises of farming communities may be significant seeing as endotoxins within the human body help to regulate the immature immune system (Riedler et al., 2001). Note that other work does suggest the reverse and that poor living and working environments directly increase the risk of skin allergies and asthma (Chandramohan & Sivasankar, 2009). International research concerned with finalising the underlying causes of this relationship between living conditions and immunity is focussing on (1) the role of hookworm and other gastrointestinal infections in reducing the degree of skin sensitivity, possibly by influencing T-cell responses in very young children (Flohr et al., 2006), and (2) the impact of consumption on intestinal microflora (for example, from antibiotics) and the suppression of intestinal microflora, which are vital for developing the immune system of young children (Flöistrup et al., 2006) or from consuming non-infectious microbial components, for example from raw cow milk (Riedler et al., 2001). This is clearly not to say that low socio-economic status offers protection from all skin diseases. Poor living and working conditions increase exposure to other potentially lethal diseases, such as epidermal parasites (e.g. scabies and lice) which disproportionately affect impoverished communities (Feldmeier & Heukelbach, 2009). The reality of the situation is clearly very complicated, as is demonstrated by the study of 102,353 American children which found a positive relationship between wealth and parental education levels and skin and respiratory allergies but a negative relationship with other infections (Victorino & Gauthier, 2009).

This notion of our social environment and circumstances determining our biological functioning will be explored repeatedly in other areas throughout this volume. Skin colour and socio-economic status are also closely interrelated and these can have significant health outcomes. In the next section we explore the role of ancestry and skin colour on identity and identification.

3.3.4   *Ancestry*

> To be black is to be not-white; to be white is to be not-black. These colour terms are not in fact chromatic; but algebraic. (Connor, 2004: 148)

Skin colour has played the central role in almost all categorisations of race to date (Relethford, 2009). By the seventeenth century, skin colour became one of the primary characteristics for ethnic differentiation (Benthien, 2002). By the eighteenth century, skin colour took on a new, more sinister dimension as a key basis for discrimination in theories of 'racial difference' (Connor, 2004: 147). White was viewed as the basic natural state of man and numerous attempts were made to depigment black skin (Benthien, 2002). Crude racial categorisations based on skin colour are, of course, still widely utilised today, particularly in folk taxonomies (Mukhopadhyay & Moses, 1997). Such is the power of skin colour as a visual cue for 'otherness' that men and women of differing ethnic groups have been described as 'imprisoned by discourses of skin' (Tate, 2001: 209). First we discuss the biology of skin colour before assessing the interaction between colour, environment and culture.

Skin colouring is highly heritable and is argued to involve a polygenic combination of traits influenced by a range of single nucleotide polymorphisms (van Daal, 2008). Human pigmentation derives mostly from melanin (a photoprotective feature) within the skin. The degree of pigmentation results from the production and deposition of melanin as either black/brown eumelanin or yellow/red pheomelanin, with higher levels of eumelanin characterising the darker skin phenotypes (Jablonski, 2004; van Daal, 2008). Melanin levels may be constitutive skin colour (genetically determined) or facultative skin colour ('tan' as a result of solar radiation). Once formed, the melanin is located within the keratinocytes with the subsequent release of melanosomes within the keratinocyte cytoplasm resulting in colour and protection from the UV light; a failure to effectively synthesise melanin results in oculocutaneous albinism (van Daal, 2008).

While 'skin colour is almost synonymous with "race"', it is in reality under strong selective pressure (Brace, 1995: 172). Similar skin colour can relate to convergent adaptation as oppose to genetic relatedness. As discussed earlier, anthropologists have a long history of categorising people with different skin colours and numerous descriptive scales and charts have been produced and abandoned (e.g. von Luschen, 1897). As with all such categorisations these mask the reality of the continuous range of skin colours from fair to dark (Relethford, 2009). Also, they fail to recognise the fact that different cultures perceive colour differently (see Chapter 4). Skin pigmentation charts have mapped the skin colour of indigenous peoples geographically and show a preponderance of darkly pigmented populations in the southern hemisphere and near the equator, but a predominance of lighter skin pigments closer to the poles. Ultimately there is a strong correlation between skin pigmentation and latitude that is related directly to sunlight (Aoki, 2002). In modern populations, such maps are problematic due to enormous amounts of mobility and migration. Research has shown that early hominids were all darkly pigmented (Jablonski & Chaplin, 2000). Ultraviolet radiation intensity has been one of the primary selective factors with regard to skin pigmentation; diet and culture also play a part. Depigmentation of the skin is thought to have evolved when humans spread beyond the tropics in order to maintain sufficient levels of vitamin D3 in the face of diminished exposure to sunlight (UVB wavelengths). The potential impact of cultural practice on this overall pattern can be observed, however, in the skin colour of Inuit groups whose skin is darker than their latitude would predict. This is related to their predominantly marine diet, which is rich in vitamin D3, much of which is stored in the skin as fat (Jablonski, 2004). It is argued that there is a selective pressure for darker skin in latitudes closer to the equator, in those areas where UVB levels are consistent and strong, though this effect can be mitigated by diet (Chaplin & Jablinski, 2009; Juzeniene et al., 2009) (though see Robins, 2009 for a counterargument and response by Chaplin & Jablonski, 2009).

It is ironic that skin colour is one of the strongest phenotypes through which people are grouped ethnically, yet it is an adaptive phenotype that has been described as extremely labile and provides greater evidence concerning shared past environments than genetic ancestry (Jablonski, 2004; Klimentidis, 2009).

The ability of humans to alter skin colours in order to adapt to differing environments has been an important aspect of the evolution and the migration of our species (Juzeniene et al., 2009). Because skin pigmentation may be associated with latitude, but not longitude, skin colour variation is of limited use in the identification of geographical origin (Brace, 1995). As Brace discusses, individuals from as widely dispersed regions as Australia, Africa and India are grouped together based on nothing more than the degree of skin pigmentation: 'the "something there" is melanin, but that is all' (Brace, 1995: 171). The importance that skin colour has as a marker of ethnicity seems nonsensical given its spurious basis for categorising people:

> The apparent existence of a difference between so-called human races and subgroups is predicated on an exaggerated perception and heightened sensitivity to a visually obvious attribute of human appearance. (Jablonski, 2004: 615)

Ahmed writes: 'One's racial identity is not simply determined by the "fact" of one's skin colour. Racialisation is a process that takes place in time and space: "race" is an effect of this process rather than the origin or cause. In terms of skin colour "racialisation" is the process of investing that skin colour with meaning such as "black and white". These come to function not as descriptors of skin colour but as racial identities' (2002: 46) (see quote by Connor, 2004). Race as bodily difference is a consequence of rather than an origin of ethnicity.

The way in which race or ethnicity is constructed within a society and the way in which it is acted out has biological implications. Dressler et al. have discussed how race and ethnicity are the most commonly used variables in public health research, though these terms are used uncritically and without any standardisation or definition (2005). Only relatively recently have social factors related to race and ethnicity been properly considered in studies of ethnicity and health. Institutional racism (e.g. job and housing inequalities) and perceived racism (self-reported experience of discrimination in day-to-day life) have been evaluated in this psychosocial model (Dressler et al., 2005). Skin colour and the meaning differing societies invest in it means that it is the direct cause of social inequalities and psychosocial pressures associated with racism. These, in turn, are the source of considerable health inequalities. Studies of the association between skin colour and health have measured skin colour using reflective spectrophotometry, self-rated skin colour and ascribed skin colour. In several studies, ascribed skin colour (as opposed to that determined through measurement of skin pigmentation) has been correlated with high blood pressure when all other socio-economic factors have been controlled for (e.g. Gravlee & Dressler, 2005; Gravlee et al., 2005 for Puerto Rico; Dressler et al., 2005; Sweet et al., 2007). Not only is there a difference between black and white Americans with regards to hypertension, but there is a gradient according to skin colour within the African American community itself. It is significant that it should be self-rated and culturally ascribed colour, rather than actual skin pigmentation as measured objectively using photo-sensitive devices, that has this association. Other studies have shown that within some societies, those with lighter skin tones have a higher socio-economic status (e.g. López, 2008). However, López found that a greater attachment to a particular ethnic identity resulted in higher self-esteem, compared to those with lighter skin tones but who felt less culturally embedded (2010). Indeed, extended kin, strong cultural identity and friendships (cultural consonance) have been found to buffer the effects of other psychosocial stressors (Dressler et al., 2005; López, 2008).

The Western world has invested melanin with enormous social meaning, constructing differences among people that have had very real biological consequences with respect to health outcomes. In terms of human identification, the degree of skin pigmentation has little genetic basis and provides extremely limited information concerning an individual's ancestry. In terms of human identity, it can impact gender relations, social stratification, cultural identity and health. The skin's outer surface acts as a protective barrier to the inner world, but with regards to colour, it has the potential to be invested with meanings of 'otherness', becoming the focus of hostility and resultant psychosocial stress.

3.3.5   *Defective skin*

So far, we have discussed the relationship between life, identity and healthy skin; the skin, however, has been an important tool for the diagnosis of disease for millennia (Connor, 2004). A person's pallor along with blotches, spots, pustules and boils, all display inner malaise. Dramatic diseases affecting the skin include leprosy and syphilis, both of which are enormously stigmatising for the sufferers (Wilson, 1999). Leprosy in medieval England was believed to be a moral and spiritual contagion as much as a physical condition – inner depravity reflected by the corruption of the outer surface, that is to say, you and your inner identity reflects through your skin (Rawcliffe, 2006). Diseases specifically affecting the skin are often not life-threatening (e.g. psoriasis) but have significant impact on the quality of life for the sufferer and are accompanied by considerable social stigma. As a consequence, many of those suffering skin diseases experience greater psychosocial distress. In one study of individuals with skin conditions including psoriasis and acne, the majority reported feeling shame and embarrassment, with many stating that the worse thing about their condition was the way that it looked: 'It looks like the plague' (interviewee quoted in Jowett & Ryan, 1985: 426). Interviewees expressed a concern that other people might think the condition stemmed from a lack of hygiene. The presence of these ailments curtailed the sufferers' engagement with particular sporting activities and also from pursuing certain careers, all because of the appearance of their skin (Orton, 1981; Jowett & Ryan, 1985). Skin diseases are common and represent a high proportion of appointments to general practitioners (Orton, 1981; Picardi & Pasquini, 2007). They can present in a range of ways, including alterations of skin colour, itching, thickening and pain. Sufferers of skin conditions also report bullying, even with relatively common skin disorders. The skin is sensitive to not only physical condition, but also mental state. Stress-related eczema, for example, is very common. Connor describes a particularly poignant example of this, from the daughter of a Holocaust survivor whose guilt at this survival manifested as a patch of eczema in the place where her mother had her camp tattoos (2004).

The stigmatising affect of blemished skin has a long history within the Western world. For example, moles on the face were thought to indicate bad fortune, and it was suggested in the eighteenth century that a pregnant woman's imagination could imprint her developing baby with marks and deformities (see Wilson, 1999). Skin disorders are especially stigmatising in our own air-brushed-to-perfection world. Those who fall short of the impossible ideal may endure a range of social problems. The sufferers in Orton's book describe different disorders but similar senses of alienation and despondency. Suffering from a skin disorder also links to one's sense of identity, either by being defined by your disorder (which ironically, is what happens when Orton [1981] describes her interviewees), by having the disorder determine your activities and thus aspects of your identity or by directly influencing the disorder itself (for example, through certain activities or through emotional state).

3.3.6   *Border disputes: one body = one individual*

The skin as a boundary has meant that it has been the focus of much research within the social and physical sciences (Lafrance, 2009). The sense of touch is extremely important, with Montagu going so far as to suggest that it is the most important of senses, and that 'no organism can survive very long without externally originating cutaneous stimulation' (1978: 318) – that is, it is impossible to survive without touching and being touched. The sense of touch varies on different regions of the skin, with the hairless areas having sensation spread evenly over the papillary ridges, while the hairy areas develop small spots of sensitivity (Cauna, 1954). However, since Montagu suggested that it was a woefully under-researched topic in his 1978 volume, much theoretical and practical study has been undertaken. The sense of touch is extremely important in terms of defining the self and one's own body. The significance of touch for the developing infant has been highlighted by several authors in the field of psychoanalysis (see Lafrance, 2009 for a discussion). Murray argues that our skin provides us with a 'sentient, tactile and optic bodyscape … which underpins an experience of physical individuality' (2001: 117), while, earlier, Valéry wrote that 'the skin is the meeting not just of the senses but of world and body: through the skin the world and body touch, defining their common border' (1933: 97). This notion is probably taken for granted, although the cases when confusion and ambiguity appear force us to contemplate this further. Sensation and authorship of action is used by us all to define our self and our boundaries. But this is not always the case. For example, Gallagher discusses the case of schizophrenic patients who do not claim authorship of actions, yet still feel sensation in the typical way (2000). Perhaps the skin's significance in helping to define a human individual, a sense of 'me' and 'other', can be truly seen with the experiences of conjoined twins. Here both sensation and authorship of action can be shared. As a result, a shared embodiment is seen as conjoined twins present an 'anomaly in relation to the propounded properties of bounded individuality' (Murray, 2001: 120). In essence, individuals are seen to overlap. Where their bodies connect, it is entirely possible for both conjoined twins to feel sensation and to even have joined control of limbs.

Conjoined twinning occurs as a result of an incomplete separation of the developing single embryo, and where the debate moves from the theoretical to the real is when it is decided that the conjoined twins must be forcibly separated. This is invariably undertaken when the twins are young. Key decisions must be made with regard to organ and limb ownership – even though, as has been said, each twin can claim ownership through sensation, authorship of action and function. It has been argued that the sense that certain key organs belong to one individual and not the other is 'a morally irrelevant locational explanation' (Gillon, 2001: 4). Nonetheless, the medical profession will make these decisions, potentially putting lives at risk that were not previously so, according to Murray (2001), because of the affront to our conceptions of self that conjoined twins provoke. Of interest is the reaction of newly separated twins who state that the separation equates to the loss of a limb or to a bereavement (Murray, 2001). The result is a loss of self; an erosion of identity.

In 2000, the English Court of Appeal confirmed the legality of doctors overruling parents' decision not to separate conjoined twins. The ruling argued that the doctors' decision to choose the 'least worst' scenario of allowing one child to survive at the expense of the other was entirely justified. Although many, such as Easterbrook (2001), argue in agreement of this utilitarian decision, the pragmatic reality is a little less straightforward. Gillon points out that the issue is far from clear-cut as it is not possible to state which of the two arguments presented (kill one child to save the other versus not performing a legitimised killing but thus allowing both children to die naturally) is the 'least worst': both are morally sound and permissible by law (2001). Indeed, it would be interesting to see, as Martin-McDonald and colleagues consider, if informed consent was possible from the conjoined twins whether they would agree to such a procedure (2005). Some argue that such procedures should not be conducted for moral or religious reasons, but for economic ones; that the financial burden of the operations and subsequent care of the twins cannot be justified and that the money would be better spent serving a larger number of ill people in the community (Martin-McDonald et al., 2005). This is a dangerous argument, however, as many treatments such as fertility treatments and cancer therapies are financially intensive, and one would find it difficult to deny that these were valid uses of public money. Regardless, the action of the doctors was to confirm Murray's assertions that rather than separating bodies, medical intervention surgically and socially constructs these offending bodies around our dominant embodiment paradigm of one body = one individual (2001).

3.4   Conclusion

The discussion in this chapter has highlighted the centrality of the skin for both human identity and identification. It is the largest organ of the body, and as we have seen, is surprisingly complex throughout all of its layers. It provides a vital focus for the identification sciences, and the individualising properties of characteristics such as fingerprints have been noted for thousands of years. Yet the skin continually develops and alters over the life course and becomes impregnated with our social experiences: it materialises life memories (referred to as 'dermographies') through stretch marks, scars and so forth. These memories, however, are imperfect and may be impermanent due to the skin's ability to heal and regenerate (Prosser, 2001). The skin is often referred to as a canvas, upon which a multitude of aspects of our identities are painted and subsequently interpreted. We have seen this clearly in terms of sex, age and ancestry. Yet it is a shifting canvas, one which externalises identities which may betray the image we wish to project. As Prosser suggests, within modern Western society, good skin is that which forgets, which does not mark the passage of time or record physical onslaughts such as pregnancy (2001: 54–5).

The skin has a double-sided nature, in that it both 'protects the body from the external world, yet at the same time reveals and communicates the internal state of this primal vavity to the external world' (Gell, 1993: 29). The skin has a significant role to play in our personal identities in that it mediates the world through actually mingling with it (Connor, 2004) and provides a conduit for communication between our internal and external selves/worlds (Gell, 1993). As Turner argues, 'the surface of the body becomes, in any human society, a boundary of peculiar complexity … which simultaneously separates domains lying on either side of it and conflates different levels of social, individual, and intra-psychic meaning' (1980: 139). In the following chapter we penetrate the skin's protective barrier in order to examine what lies beneath.

4

Blood and guts

Everything seems so lovingly packaged and arranged, like a cabin trunk stowed against breakage with just those items necessary for the voyage.

(Dibdin, 1992: 99)

While the skin provides the outward-facing surface and is, therefore, a greater contributor to external interpretation of our bodies, many components of our hidden internal bodies are also significant in terms of our identity. From horror films to medical exhibitions, inner organs, blood and guts have been exposed to the general public as never before. And yet the sight of these structures continues to provoke strong reactions – particularly when displaced from the body's familiar container of skin. Advances in medical technology such as CT and MRI scans now mean that the internal workings of the human body can be visualised without the need to transgress or pierce the skin. Thus identification of the living is no longer confined to their external phenotype. New biometric and imaging techniques for the identification of individuals have been developed due to the desire to defend against threats to security, whether on a personal, corporate or national level. Here we discuss some of the internal structures, substances and soft tissues most frequently harnessed for the purpose of human identification, including the vascular system, blood, eyes and body fat. We discuss each of these in relation to their perceived connection to social and group identity. In the final section, we consider the transplantation of organs and its impact on human identification analysis as well as on the identity of both the donor and the recipient.

4.1   The vascular system

The importance of the vascular system is in maintaining homeostasis through its network of arteries, veins and capillaries (Pugsley & Tabrizchi,

2000). Although we tend to consider it on its own, it is better to think of it as paired with the lymphatic components (Pugsley & Tabrizchi, 2000) which remove waste products from around the body. As is well established, blood flows from the heart to the lungs to be oxygenated and then is pumped around the body in arteries, arterioles and capillaries and then back to the heart via venules and veins. In reality, this is a highly complicated system; we must remember that it occurs in three dimensions and the blood vessels range widely in diameter (Guidolin et al., 2011). In fact, without the complexity of this system, higher-level animal species could not have evolved (Plante, 2003). The structure of the arteries and veins is similar but the different ratios of smooth muscle, elastic fibres and collagen results in thicker and stronger arteries (Pugsley & Tabrizchi, 2000). Further, the branches and angles of the bifurcating vascular system are precisely located and are influenced greatly by the force of the flowing blood – to the extent that a lack of flow can cause resorbtion of the vessels (Guidolin et al., 2011). The cardiovascular system is actually the first system to develop in the vertebrate embryo (Pelster, 2003; Guidolin et al., 2011), although in some animals initial blood flow is not actually linked to the metabolic requirements of the organism (Pelster, 2003). They form either through a process of vasculogenesis (the formation of capillaries from endothelial cells) or, later, angiogenesis (the formation of capillaries from pre-existing vessels) (Guidolin et al., 2011).

The significance of the cardiovascular system has been known for hundreds of years. In their review of its history, Loukas et al. note that the heart was thought to be at the centre of the cardiovascular system from as early as ~2900 BC in Egypt, the concept of circulation is recorded in Chinese manuscripts from ~2689–2599 BC and evidence for an understanding of the separate arterial and venous systems appears in Greece around 500–450 BC (2007). As an aside, and a reflection of early concepts of medicine and anatomy, the word 'artery' derives from the ancient Greek *arteria* or windpipe, as arteries were thought of as filled with air (Bardell, 1978; Loukas et al., 2007). Galen writes in the second century AD that arteries breathe through the skin (Boylan, 2007). While the importance of blood for the 'nourishment' of the body has long been known, only since the seventeenth century, when William Harvey published his findings, has the circulatory system, the connection between the arteries and the veins, become better known (Poynter, 1957; Bardell, 1978).

A number of published studies describe the considerable variation among individuals in the location of their soft tissue structures, in particular neurovascular bundles (Wang et al., 2008). These studies have been both descriptive and biometric in nature, and derive primarily from a clinical perspective, as such variation is an important consideration during surgery (Olave et al., 2001). Since researchers widely acknowledge that vascular positioning varies in the human body, it is unsurprising that this variation has now been harnessed as a tool for human identification and identity verification. Tower was an early researcher in the vascular patterning of the eye, though he specifically stated that one could not necessarily extrapolate uniqueness in one vascular region or another (1955). Differences in vascular patterning among individuals underlie the identification techniques already used in retinal scans. The use of vascular trees in the limbs has also been proposed, but to date the degree of variation within a population is still not known quantitatively or statistically. The stability of these structures over an individual's lifetime is also not fully known, and it has been shown that ageing affects all regions of the cardiovascular system (Plante, 2003).

Despite these reservations, vascular recognition technology has developed and been widely trialled and implemented in a number of different identity authentification contexts (Lopes, 2010). For example, finger or palm vein biometrics are widely used as a source of identity verification in banks in Japan – the client registers his or her vein pattern on a smart card (Elliott, 2011). Infrared scanners read the hand and verification is achieved when the vascular pattern matches that stored for the individual. Vascular patterns or vein technology is seen to have an advantage over fingerprinting because it is internal and leaves no trace. Fingerprints by contrast can be left on surfaces and thus are susceptible to being stolen. An additional feature of palm-scanning equipment is that it can also include a 'proof of life', that is, it can read a pulse. In the case of fingerprint security, it is possible in some contexts that this can be breached through the use of the relevant person's finger even if he or she is dead. Additionally, new palm-scanning equipment is believed to be quicker and more reliable than the equivalent fingerprint technology (Lopes, 2010). Finally, vascular scanning generally does not have the same social stigma attached to it as fingerprinting – people are likely to be more compliant because the act of observing a vascular pattern does not have the same criminal associations. Given these advantages, it is likely that vascular pattern techniques will become more widespread in the future.

On a social level there is a close correlation between blood pressure, cardiovascular disease (CVD) and ethnicity and social class. For example, it has been noted that both self-rated and culturally ascribed colour – but not actual skin pigmentation – are associated with blood pressure, likely through an interaction with income and education (Gravlee & Dressler, 2005; Gravlee et al., 2005) (see Chapters 2 and 3).

4.2   Blood

The content of the vascular system, the blood, is a highly emotive substance and almost universally a symbol of life (Scheper-Hughes & Lock, 1987). Blood is the subject of mythologies, hate, religious fervour, pollution – it provokes both strong negative and positive reactions. It has a powerful resonance with self and group identity and is closely enmeshed within theological beliefs: 'Blood is theology transformed' (Anidjar, 2005: 120). Discourses on ethnicity frequently invoke 'blood' to represent kinship: 'blood is thicker than water'. In the Middle Ages and in early modern England, blood was central to beliefs concerning life, health, relationships and procreation (Crawford, 2004; Bildhauer, 2006). While the notion of 'blood relations' or 'blood ties' remains symbolically potent in today's society, in the Middle Ages blood was actually believed to be passed on to children, who were literally viewed to be 'of the same blood'. The term 'bad blood' is still sometimes used to explain criminal behaviour in those who are related. Thus to share common blood was significant for drawing together kin as one body, particularly in the face of outside threats (Crawford, 2004; Bildhauer, 2006). In medieval medical thought, the key to understanding the cause of ill health was to understand the state of the blood (Bildhauer, 2006: 23). While the body contained the four humours – blood, phlegm, black bile, yellow bile – blood was the most important because it transported the other three (Bildhauer, 2006: 24). Blood was also believed to be the bodily fluid that produced sperm and maternal milk (Crawford, 2004). Blood features particularly strongly in medieval literature, and in Bildhauer's fascinating book about blood in the Middle Ages she writes:

> References to blood were often used to make a truth claim forceful enough to count as absolute as incontrovertible proof … Again and again, blood furnished such authentification not only of Christ's, but also of any human body. (2006)

While blood and kinship are strongly linked, so too is blood and ethnic identity. In the past, censuses dealing with race in the United States used a 'one drop' policy, under which one drop of 'negro blood' meant a person was classified as black. Blood featured prominently in racist discourses of the nineteenth century, which attempted to show that race went more than skin deep. As Epstein discusses, according to a medical journal published in 1851, the black person's 'bile … his blood … the brain and nerves, the chyle and all the humours' were all 'tinctured with a shade of the pervading darkness' (2004: 190). Furthermore, blacks, it was argued, suffered from a lack of red blood, a condition which, fortuitously for the period, could be cured through hard exercise – thus neatly justifying slavery (Tucker, 1994: 13–14, cited in Epstein, 2004: 190). Ironically, it was the study of the blood groups that first helped to dismantle the racial world view concerning bounded, homogenous categorisations of individuals.

Today blood continues to have a strong symbolic resonance (see Carsten, 2011 for a review) and it has significant information to offer in terms of identity and identification. The anatomical properties and components of blood are described by many general publications, including Ferenc (2008) and Gunn (2009). First, it is important to emphasise that the term 'blood' refers to both the liquid and the solids suspended in it. Around 70 ml of blood is present per kilogram of body weight in an average person. The solid component is not insubstantial, comprising around 45 per cent of the volume of blood. This component includes red and white blood cells and platelets derived from precursor stem cells from the red bone marrow. The red blood cells, or erythrocytes, are approximately seven to eight micrometres in diameter and about two micrometres in thickness and work to carry oxygen around the body. These biconcaved discs contain no nucleus and survive for up to 120 days before being removed by the spleen or liver. Males and females have different quantities of red blood cells with men having between $4.5–5.9 \times 10^6$ per microlitre and women between $4.5–5.1 \times 10^6$ per microlitre. White blood cells help regulate the immune system and fight infection. They come in a variety of forms, such as lymphocytes and granulocytes, but are nucleated and thus have the potential to provide DNA for analysis. A healthy individual will have around $4.4–11 \times 10^3$ per microlitre of blood. Platelets, or thrombocytes, are like red blood cells and lack a nucleus, but they are smaller in size (around two to four micrometres) and survive for only about nine to twelve days. The role of platelets is to stop blood loss (hemostasis) by forming a lump around a site of leakage and simultaneously releasing trigger chemicals to permit and encourage other clotting processes. The average number for a healthy person is one hundred fifty thousand to four hundred fifty thousand per microlitre of blood, and spontaneous bleeding will occur if this number falls below twenty thousand per microlitre. All of these solids permeate through the liquid phase of the blood, known as plasma. The plasma is viscous, straw coloured and composed of 90 per cent water; the remainder being salts, proteins, lipids, dissolved gasses and so forth. Although blood is clearly vital for existence, some of its components are actually harmful; for example haemoglobin when removed from its red blood cell container is highly toxic (Knuti et al., 2002; Newton, 2007).

Not all blood is the same, and differences in the structure of the key elements have important implications for identity and identification. Over

100 years ago, Karl Landsteiner described the ABO system of blood classification and identified its potential for human identification (Goodwin et al., 2007). The familiar ABO classification system is based on the fact that while certain antigen structures found on the red blood cells contain glucose, galactose, N-acetylglucosamine and fructose (the O variety), there is the possibility of an extra N-acetylglucosamine molecule attached to the galactose (the A variety) or an extra galactose molecule attached to the galactose (the B variety) (Newton, 2007). This has implications for organ transplantation and the management of individuals with illnesses. Rejection of transplanted blood and organs is likely if the antigen structures of the donor do not match those of the recipient. In essence, the body does not recognise the inserted tissues and treats them as a threat. In terms of organ transplantation, reviews of case work have shown that identical matches are not always necessary, but that ABO-compatible tissues will suffice (Neves et al., 2009). Although the use of compatible rather than identical tissues poses a slightly greater risk of rejection, it does drastically reduce the waiting time for patients with less common blood types (Neves et al., 2009).

In terms of human identification analysis, in some contexts the blood may not be any longer contained inside the individual. Analysis of blood stains in the absence of a body is therefore a significant component of forensic work. In such crime scene contexts, one of the first challenges may be to try and detect the presence of blood in the first place. A particular challenge for those working in the serological arena is the differentiation of blood stains from other reddish stains. This is also complicated by the physical changes that blood cells undergo when drying in the atmosphere. First, the red blood cells shrink and lose their concave nature by becoming rounded with scalloped or spiky surfaces. Continued exposure will then cause their membranes to rupture and split (hemolysis) (Ferenc, 2008). A range of presumptive tests must therefore be applied. First, an alternative light source can be used to highlight certain stains – although it should be noted that certain light wavelengths can destroy DNA (Virkler & Lednev, 2009). Second, presumptive tests can be deployed which, as Newton describes, work on the principle of applying a substance to the stain which will react chemically with the contents of blood, usually the haemoglobin (2007). No reaction, therefore, means that the suspect stain is not blood.

A key problem with many presumptive tests is simply that one must make an assumption as to the presence of blood before applying the appropriate test to confirm whether it is there (Virkler & Lednev, 2009). Although different products are available on the market, many common methods involve the exploitation of the heme catalyzing with a dye and peroxide. Here, the haemoglobin in the blood acts in a peroxidase-like manner, catalyzing the oxidation by peroxide of the dyes which results in a change of colour (Marie, 2008). Luminol, often referred to in television programmes, functions slightly differently in that it is 5-amino-2,3-dihydro-1,4-phthalazinedione, which when in the presence of blood and hydrogen peroxide reacts to release nitrogen gas and 3-aminophthlate, which in turn releases a photon which fluoresces bluey-violet (Newton, 2007). As such, it can be of limited use in some environments (Virkler & Lednev, 2009). This reaction only occurs once, so subsequent fluorescence of a given blood stain will require further sprays of luminol (Marie, 2008). The universal adoption of luminol is partly due to its very high sensitivity (it can detect blood diluted 10 million times) and its success with stains even a few years old (Newton, 2007). Marie warns, however, that these are merely presumptive tests and further analysis is required (e.g. using genetic techniques) to confirm definitively that these are blood stains and from humans (2008). Emerging techniques tend to focus on the use of infrared, X-ray fluorescence or immunological protocols for their success (Virkler & Lednev, 2009).

Attempts have also been made to detect blood from the archaeological record, and this provides a nice example of the controversy that can surround such work. The detection of blood from ancient contexts can potentially reveal information pertaining to human evolution, disease predispositions of populations, subsistence patterns and animal migration. A number of attempts in the 1990s were made to extract and analyse blood residues from within human and faunal skeletal assemblages in addition to the tools used to hunt and kill (Cattaneo et al., 1990, 1995; Kooyman et al., 1992; Smith & Wilson, 1992, 2001; for a discussion see Brown & Brown, 2011). One method of examining ancient blood is to study the shape of the crystals of the haemoglobin on the artefact of interest. The shape, length and angles of these haemoglobin crystals vary from species to species, and in theory such studies have the potential to achieve an extra layer of information (Bahn, 1987; Smith & Wilson, 1992). A key problem with this technique is that the crystals under examination tend to be deformed or damaged by the depositional environment over time (Smith & Wilson, 1992; Brown & Brown, 2011). Unfortunately, laboratory studies have demonstrated that even when pure human haemoglobin has been prepared and analysed, a wide range of crystal shapes may be present (Smith & Wilson, 2001). Much of the research on blood residue analysis in the archaeological record has proven highly controversial in terms of the veracity of the results and has elicited strong debates in the academic literature. Brown and Brown provide a very good summary of these debates (2011). Some authors strongly advise a degree of caution when interpreting all results of the detection of

Table 4.1 *Frequency of blood types in the general population*

Blood type	Frequency in population (%)	Rhesus group	Combined frequency
A	40–42	+	1 in 3
		–	1 in 16
B	10–12	+	1 in 12
		–	1 in 67
AB	3–5	+	1 in 29
		–	1 in 167
O	43–45	+	1 in 3
		–	1 in 15

Source: Adapted from Newton, 2007

ancient blood; specifically they note that little authentication has been undertaken in the discipline and thus we should adopt a healthy scepticism (Smith & Wilson, 2001; Brown & Brown, 2011).

4.2.1   Blood and identification

Although blood is now used less frequently for identification purposes, as it has been superseded by DNA profiling (Chapter 6), the use of blood typing still has its place in forensic science, arguably more so in regions with access to less sophisticated forensic and analytical apparatus. The concept is straightforward: blood typing cannot identify a person uniquely, but it can narrow down a list of suspects by declaring whether an individual and a blood sample contain the same general types. The two forms of blood grouping best known for their forensic application are those of ABO and Rhesus +/– classification. Table 4.1 shows the probabilities of these commonly used blood types in the general population, and thus shows how they can help include or dismiss individuals from investigations. Other classification systems exist and the combination of several can improve the probability of finding a subject.

In addition to the red blood cells, investigators can use the contents of the blood plasma for identifying individuals. Polymorphic proteins and isoenzymes – proteins and enzymes that perform the same biological function but differ in subtle structural ways – can be used in a similar way to blood types in terms of comparison of sample material with frequency of occurrence in the population, but their significance has lessened with the spread of methodologies related to the Polymerase Chain Reaction (Hochmeister, 1995).

In modern medicine, oxygen-carrying alternatives (either Perfluorocarbon or haemoglobin based) are used for blood transfusions. It would be interesting to see the effects of the integration of these into the bloodstream in terms of human identity and identification. Regarding identification, there are interesting cases of 'chimera', a phenomenon whereby a single organism contains the genetic material originating from two different zygotes. One might presume that the occurrence would be so rare as to be extremely useful in confirming identification. One might also expect an alteration in individuals' perception of their own bodies, seeing as they may be incorporating animal or artificial components into their bodies. Similar experiences must be had by those undergoing such procedures as heart valve replacement surgery and organ transplants (Sharp, 2006). Although successful oxygen-carrying alternatives have been slow in coming (previous attempts have resulted in acute cardiovascular, renal and pancreatic and coagulation toxicities), Hemopure® holds great potential. Hemopure® is an ultrapurified, glutaraldehyde-polymerized bovine haemoglobin in a balanced electrolyte solution; it has a similar molecular structure to human haemoglobin but is more efficient at carrying oxygen (Sprung et al., 2002). A small study of forty-two patients by Sprung and colleagues noted that the use of Hemopure® caused no serious adverse events and that haemoglobin and platelet measurements and blood chemistry were normal; only mild skin discolouration and increases in blood pressure and heart rate were experienced (2002).

Acts of trauma and violence can also lead an investigator to the examination of blood for identification purposes. Indeed, blood is thought to be the body fluid most commonly encountered at a crime scene (Virkler & Lednev, 2009). Although the analysis of blood spatter patterns cannot confirm either the victim or the attacker's identity, it can provide a wealth of information regarding the context and manner of the assault or death. Examples of the sort of information that can be gathered include the class of weapon used or the orientation of the attacker to the victim (Karger et al., 2008), but this assumes that the blood has been forced out of the confines of the body. Even when this is not the case and the skin has not been ruptured, the way in which blood collects within the body can provide useful information. Traumatic bruises and contusions allow the investigator to make inferences about: the weapon used to cause the damage, from the shape and outline left on the skin; the period of time since injury due to the colour change of the bruise; the time since death, from the degree of *livor mortis*, the post-mortem settling and decomposition of blood in the body; and the position of the body at death, from the position of the blood during *livor mortis* (Ferenc, 2008). This falls outside the scope of this discussion, however, and interested readers are directed to the likes of Bevel and Gardner (2008) and Wonder (2007) for greater information and contextualisation.

4.2.2   Blood and identity

Blood can have an identity even outside of the body. People have an intrinsically strong reaction to the sight of blood and touching another person's blood is thought to be highly polluting, associated as it is with the transmission of disease (particularly since the 1980s and HIV awareness campaigns). Cross-culturally, menstrual blood is usually something to keep hidden and to attend to, and associated feminine products should be kept discreetly out of view – particularly in male company. While blood transfusions occur within the context of the medical setting and are subject to the same objective clinical language that suffuses all practices within this arena, they can also be highly emotive. To have the blood of another person within you may be perceived as 'polluting' by some cultural and religious groups (Strathern, 2009). As Lambert and McDonald observe: 'human matter when divorced from its source is frequently a focus of concern and elaboration' (2009, 3).

Our blood has interesting, though underexplored, implications for our identity (see Copeman, 2009). For example, blood-based pathologies can become an intrinsic part of a person's identity, having a significant impact on their lives. Some can even affect the sufferer's physical appearance, making the affliction visible to others and further influencing a sense of identity. Haemophilia, a condition in which the clotting mechanism is faulty or absent, is one such pathology (Ferenc, 2008). Genetic anaemia such as sickle cell anaemia and thalassaemia are also of interest in terms of identity and identification. These diseases tend to be restricted to individuals living in parts of the world where malaria is still rife as they confer a genetic resistance to this disease. In archaeological studies of skeletal remains, the presence of evidence of this condition in non-malarious countries has been used to infer mobility. For example, Lewis (early view) describes the skeletal remains of a child with pathological abnormalities consistent with thalassaemia from a Roman site in England. This evidence, though indirect, suggests population mobility – the child's parents likely travelled to England – given the genetic affinity with populations originating from the Mediterranean.

Beyond this, some interesting recent thinking has suggested that the contents of blood may have a very real impact on a person's appearance and even sexuality. Researchers have argued for a link between certain non-standard phenotypes such as hair whorl orientation, handedness, speech laterality and sexual orientation. In a controversial publication, Hatfield argued that evidence supports the notion that the Rhesus (Rh) genes may be involved in determining whether an individual is heterosexual or homosexual (2006). He argues that previous work has already shown the influence of the Rh gene system on non-right handedness, schizophrenia, autism and speech disorders, and proposes the random recessive model to explain why those relationships, in addition to that with sexuality, have not been strongly correlated with any one gene. The random recessive model states that there is only a 50 per cent chance of the recessive phenotype occurring with the presence of the recessive genes, as opposed to guaranteed development. This can then result in, for example, one recessive person being right-handed with left brain speech laterality while another could be both left-handed and have left-sided brain laterality. His key point is that the controlling genes for these may be the Rhesus genes and that the association of these phenotypes with sexuality comes through the application of the maternal immunization theory. Hatfield notes that this is not to say an additional genetic basis for heterosexuality or homosexuality is not also at work (2006).

Blood and sexuality have been associated in other ways too. Blood donation and transfusions have been the source of a significant amount of academic interest over recent years (Copeman, 2009). Strong discusses the prevention in the United Kingdom of gay men who have been sexually active since 1977 from donating blood (2009). The infection of haemophiliacs with contaminated HIV positive blood has been framed in terms of 'innocent victims' versus 'guilty' parties. Strong discusses how the ineligibility of sexually active gay men to donate blood forges an identity between themselves and other excluded groups, but also infuses them with a marginalised, inferior identity in relation to this issue (2009). Strong thus refers to blood as a 'biosymbolic medium for the contemporary creation of persons and populations' (Strong, 2009: 174, author's emphasis) from which sexually active gay men have been excluded. Shao and Scoggin also discuss the 'contaminated blood' in a group of HIV positive individuals in China and how this group's members perceived themselves as linked by 'blood ties' because they had the 'same kind of blood' (2009). Pathogenic blood in this instance confers a group identity onto some of the sufferers that may be likened to family links. In a special edition of the journal Body and Society (2009) on blood donation, it becomes clear that blood and its extraction and insertion from one person to another has the power to significantly alter the identities of both the donor and the recipient. Blood transfusions are a source of anxiety and threat to the boundedness of an individual.

4.2.3   Blood and religion

As discussed earlier, religion and blood have long been closely linked (Anidjar, 2005). Imagery often refers to blood and Christ, holy relics may contain blood, religious statues may 'bleed'. Religious iconography and theology is interwoven with the symbolism of blood, quite apart from the blood shed in the name of religion. An individual's religious identity can have a significant impact on his or her relationship with blood and blood products. For example, some religious beliefs restrict the interaction of a person with blood. Jehovah's Witnesses are forbidden, at the risk of disassociation from the faith and ostracism from the community, from consuming blood. It is believed to be a sin of such magnitude that it restricts access to Heaven (Glendenning, 2002). This ruling follows a 1945 interpretation of three key passages from the Bible in Genesis, Leviticus and Acts (Kerridge et al., 1997; Knuti et al., 2002). In this interpretation, consumption refers to the oral ingestion of blood but also to intravenous consumption in the context of medicine. Adherents may carry information cards stating such a belief (Glendenning, 2002). This ruling was amended in 2000 to note that acceptance of blood by an individual would not result in the church expelling him or her from membership, but that the individual effectively did this himself or herself (Muramoto, 2001). The end result is therefore the same. This clarification is likely the result of a policy established in Bulgaria following the intervention of the European Commission of Human Rights which ensured that individuals had the right to choose blood transfusions freely without control or sanction from an organisation (Muramoto, 2001). Thus, the rules on the acceptance of blood are complicated.

Although red and white blood cells, plasma and platelets are forbidden, albumin, bone marrow, stems cells and oxygen-carrying alternatives are permitted (Knuti et al., 2002). Some techniques are permitted that allow the capture and recycling of an individual's own blood, as long as it does not stop moving (Kerridge et al., 1997). This has led to the notion of a primary and secondary blood component, where the secondary component is derived from the primary (Muramoto, 2001). Interestingly, these oxygen-carrying alternatives can include bovine haemoglobin. Concerns do exist, however, about the risk of infection by prions that evade purification methodologies (Knuti et al., 2002). Indeed, it has been argued that the spread of human-based pathogens such as hepatitis C and HIV/AIDS through transfusions has vindicated the Jehovah's Witnesses' beliefs (Jarvis & Northcott, 1987; Kerridge et al., 1997). The strength of commitment to this fundamental tenet of the Jehovah's Witness faith clashes strongly with the standard Western paradigm which surrounds medical practice. Knuti et al. present such a situation when they detail the case of a female Jehovah's Witness believer dying from cancer who would accept part of the standard treatment but not the final stage, the blood transfusion, which had a good likelihood of saving her life (2002). The nursing staff, oncologist and other caregivers described their difficulties in treating this person when they found it so hard to accept her views. This is clearly a clash of identities, that of the medic with that of the believer, which results in the generally unusual outcome of the identity of the medic becoming secondary. Thus the responsibility lies on the medic to adjust his or her usual behaviour in order to achieve a new balance of patient autonomy against beneficence. It has been noted that few people (15 per cent) change their minds regarding life-sustaining treatment as they deteriorate (Kerridge et al., 1997), and in Japan there is legal precedence for such refusals to constitute a basic human right (Ohto et al., 2009).

Reading through such personal accounts as provided by Knuti et al. (2002) and Glendenning (2002), it is clear to see that a sense of impotence on the part of the medical practitioner is usual. As they and others argue, a society that has freedom of choice must allow an individual to choose not to accept recommended medical care, though the law will at times view the situation differently. US courts have ruled that although a Jehovah's Witness may refuse life-saving treatment, the courts can overrule this if the refusal would lead to a child becoming orphaned (Knuti et al., 2002). Likewise, the courts can rule on behalf of the child if the parents refuse life-saving treatment for him or her. Interestingly, some within the faith have argued that the rulings of the Jehovah's Witnesses amount to coercion which removes the right to autonomous decision making of the individual (Muramoto, 2001). The refusal of blood can mean the adoption of a more risky and expensive life-saving clinical strategy and this can raise interesting questions as to whether those individuals should incur greater financial cost to cover the procedures. As Knuti et al. note, however, this is not the case for other personal choices, such as smoking (2002). This is not an insignificant or merely theoretical issue, since there are over 4.5 million Jehovah's Witnesses worldwide (Kerridge et al., 1997). Despite all of this, Jarvis and Northcott argue, that there tends to be a media bias when it comes to certain groups refusing blood transfusions, and scientists should conduct more research on the actual clinical outcomes of this (1987). In Japan, for example, only 4 in 541 deaths from surgical bleeding were associated with refusal of blood for religious reasons (Ohto et al., 2009), while with the use of hemodilution instead of transfusions in coronary bypass surgery only two in forty-six died (Jarvis & Northcott, 1987).

4.3   The eye

Eyes may be the 'windows into the soul', but they are also useful for establishing or verifying biological human identifications. In terms of social identity, they can also be of significance, particularly when they are defective in some way. For example, those who cannot see at all or whose vision is severely impaired may have their life and identity impacted in significant ways; glasses affect the way a person looks and may be accompanied by a social stigma. Although the human identification sciences are not primarily concerned with the role of the eye in terms of vision, it is worth noting in relation to human identity that different cultures may literally see the world in different ways. The eyes are easily deceived: we do not see truths; the brain fills in gaps and interprets based on prior experience (Enns, 2004). Optical illusions are a fun example of this. In modern Western society, vision has a high currency: seeing is believing, but as discussed later in this chapter, seeing is not passive or photographic, it is about interpretation and perception. Seeing is subject to cultural construction because 'There is no seeing that occurs prior to perception' (Enns, 2004: 12). For example, colours are perceived differently by different cultures, formulated and constructed in part by language and experience.

Identification and verification contexts have often placed great focus on the use of the eye because it is simultaneously exposed to the environment yet protected by the body (Daughman, 2003; Al-Raisi & Al-Khouri, 2008). While the eye has clear biometric implications, it is not particularly useful within crime scene contexts. In the absence of the body, neither foci of identification (the retina and the iris) is deposited at the scene. When a body is present, the eye undergoes relatively rapid post-mortem decomposition of the cornea, musculature and the soft tissue structures (Daughman, 2007).

Many publications summarise the structure and embryonic development of the eye, and the interested reader is directed towards these as useful examples (such as Gregory, 2007 or Bruce et al, 2003). In essence, the eye is a spherical fluid-filled structure containing a lens at one end and a responsive surface at the other. It is innervated by the large optic nerve. The basic principles of the functioning of the eye were noted first by Alhazen, an Arabian scholar alive between roughly AD 965 and 1038 (Gregory, 2007), although the detail Versalius achieves has influenced modern anatomy more (Versalius, 1543). The cornea is very unusual in that it has no blood supply (instead gaining its nutrient quota from the neighbouring aqueous humour) and is essentially isolated from the rest of the body, thus making it ideal for transplantation since the body's antibodies cannot reach the new organ to destroy it (Gregory, 2007). It can, however, become dry, and one key role of blinking is to ensure that the cornea remains lubricated. Blinking rarely occurs as the result of external stimulation; it is regulated by the brain and indeed will speed up under stress and slow down during periods of concentration – thereby making dry corneas a particularly prevalent problem in certain professions, such as architecture (Gregory, 2007). The aqueous humour is continuously replaced, and on average, is completely replaced every four hours (Gregory, 2007). The lens is often thought to be the key structure for refracting light onto the retina. This is not the case. Both Gregory (2007) and Bruce et al. (2003) explain that in reality the lens refracts light by only a small amount; it is the Refraction Index (a measure of the extent of light refraction) that is important. It is dependent on the consistency of the two adjacent media through which the light passes. The refractive index between the external air and the aqueous humour is high, although the refractive index between the humour and the lens is low, because the consistency of these two materials is so similar. Thus the aqueous humour has the greatest role to play in targeting light onto the retina; the lens essentially compensates for distance. Research has shown that the lens grows through the conversion of epithelial cells into fibre cells – although specifically it is the posterior lens cells that undergo this conversion (Iyengar et al., 2006). The formation of the lens also provides information into the reduction in eyesight with increasing age. Cells in the lens are laid down throughout life from the centre outwards. It is still not fully known why and how growth is stimulated and maintained at the periphery (Iyengar et al., 2006). Therefore, with increasing numbers of cells, those in the middle of the lens become increasingly isolated from the blood and nutrient supply. They then die and harden, thus losing their ability to change shape to allow for accommodation. Therefore as we age we begin to see the world through dead cells.

As Gregory notes, the colour of the iris makes little difference to its functioning so long as it is opaque enough to act as an effective aperture for the eye; thus eyes missing pigment such as in albinism have less efficient irises (2007). Indeed, the term 'iris' itself relates to its coloured appearance derived as it is from the Greek 'rainbow' (Wildes, 1997), and Andreas Vesalius was one of the first to use the term 'iris' in his monumental anatomical descriptors of the sixteenth century (e.g. *Epitome* from 1543). The anterior surface of the iris can be separated into two zones, the central pupillary and the peripheral cilliary areas. The border between these two zones is called the collarette (Wildes, 1997). A retinal image is not 'seen' as an object is seen – the retina is an interface (Gregory, 2007). Thus, Gregory argues, it does not matter that the retinal image is inverted since the significant factors are the relationships between the visual and tactile signals in the brain. He also notes that the retina can actually be thought of as an outgrowth of the brain's surface, and, indeed, it contains typical brain cells within its functioning structure. The structure of the retina is essentially back to front, in that the light must first travel through blood vessels and other supportive structures before reaching the receptors at the back of the retina. The optic nerve, rather than connecting directly to the receptor cells, actually connects to the supportive structures in front of those cells. The receptive cells are aligned parallel to the direction of light, with the rods detecting light and the cones detecting colour.

Underwood and Batt examine the role of the eye in reading and argue that reading is essentially the extraction of meaning from written and printed marks, which leads to information and knowledge (1996). This can be expanded, however, into the realm of identity since much culturally imbued meaning, information and knowledge in many developed societies continues to be gathered from the process of reading. Thus the role of the eye in this context is vital for the construction, maintenance or adaptation of social identity. This holds true both in terms of how one views oneself and how one is viewed by others (indeed many identity judgements and assumptions are made regarding the newspapers, periodicals or books that another is seen to read). Since the use of the eye to read is an important factor in personal identity, it is interesting to note the occasions that this may be hindered, often by wayward biology. Dyslexia, for example, is a dysfunction within the brain that makes reading and interpretation of the written word difficult and it comes in both a natural and acquired form (Underwood & Batt, 1996). Being labelled as dyslexic may also be significant in terms of one's identity. Beyond an inability to process visual information, the lack of sight is highly significant in terms of establishing an identity in the world. Lederman et al. have shown that a lack of sight does not completely negate the ability to identify a person, and their work has demonstrated that identification occurs in a more bimodal means, in which touch and sight are used in conjunction (2007).

Gregory also discusses the impact of culture upon the interpretation of information gathered from the eyes (2007). As examples, he notes how some indigenous populations live in environments with few straight lines and corners, and those living in dense jungle have difficulties perceiving depth, while conversely those living in a Western environment suffer illusions when looking down from height due to the presence of so many straight lines in our living environment. So, as Bruce et al. note, visual perception is not about seeing light, but is about seeing objects and events (2003). In this way, they note the limitations of comparing the eye to a camera. The camera produces a picture which is viewed by people, whereas the purpose of the eye and the brain is to extract information from the surrounding environment in order to guide actions. This can further be seen in the way that visual agnosia can develop following some forms of brain damage. Farah discusses this in far greater detail, but in essence this is the failure of perception and recognition of objects and demonstrates a wide array of manifestations (2004). The association with brain damage indicates the fact that the eye and brain do more than merely create a picture of the outside world. The association between the camera and the eye has interesting implications when we think of eyewitness accounts and how their unreliability is mitigated in the courtroom through the use of CCTV, mobile phone and other images.

4.3.1   The eye and identification

Research concerning the examination of the retina as a means of human identification using the eye began in the 1930s (Simon & Goldstein, 1935; Hill, 1999). Fundamentally this process works by comparing the pattern of blood vessels on the retina with a previously inputted image. Indeed, the term 'retina' derives from *rete*, meaning 'a net'. A matching vascular tree confirms identity, although in reality the process involves matching fragments of the vascular tree. Although both arteries and veins can be used, veins are preferred since they are clearer following imaging (Simon & Goldstein, 1935). Crucially, the patterns on the retina stay stable over time and are unique to each individual (Marino et al., 2006). Early work by Tower demonstrated that monozygotic (identical) twins had different and unique retinal patterns, although general similarities were present (1955). It is worth noting that the sample size was very small (six pairs of twins). Stability issues do arise following the development of retinal pathologies such as leakages (Pinz et al., 1998) or due to diabetes (Zana & Klein, 1999). These will mask the vessels of interest and therefore can fundamentally affect the ability to match a recent retinal image with a stored image. If these pathologies are stable, however, subsequent to their appearance and development, then these could prove suitably unique. In addition, Hill suggests that the low resolution of the imaging modalities may actually mask and negate small degrees of pathological change (1999). One of its key advantages is the sheer difficulty in forging such an anatomical feature, giving the technique robusticity in the field (Marino et al., 2006). Indeed, some argue for it being the most secure biometric method of identification (Hill, 1999). Another advantage of this method is that the technology for viewing the retina is readily accepted now and currently comprises part of the arsenal of techniques used as part of a standard eye examination. Research has shown that even children as young as four are comfortable with this method of examination (Hill, 1999).

Key difficulties of retinal identification concern matching vascular trees. Simon and Goldstein publicised a metric approach utilising a small grid supplemented with angle measurements in their early paper on the potential of this method (1935). More recently, Marino et al. adopted a geometric approach which, following image acquisition, attempts to match significant lengths of the vessel branches (2006). Pinz et al. adopt a similar approach, although they also utilised other anatomical features on the retina including the size, shape and location of the optic disc (1998). Other approaches seek to match bifurcation points on the vessel tree (as in Zana & Klein, 1999), in a similar way to the use of fingerprints. As with many aspects of identification, Pinz et al. suggest adoption of a method that matches different kinds of features on the retina (1998). Ultraviolet or infrared lighting are not required, although evidence shows that better definition of the vessel tree is possible with the use of green lighting or the application of fluorescein dye (Zana & Klein, 1999). More minor complications include the perceived threat to health and vision in the user and severe astigmatism in the user, making eye alignment difficult (Hill, 1999).

Once fine-tuned, an automated approach to retinal scanning can achieve a conclusion in less than half a second (see, for example, Marino et al., 2006). This is significant in terms of its acceptance and use in the real world and away from the laboratory. This fine-tuning does include tackling the issues of noise within the retinal image, blurring of the image due to eye movement, the problems associated with automatically detecting and interpreting vessel bifurcations, the fact that many of the blood vessels of interest are very narrow, the challenges to creating a system that copes with outdoor and indoor lighting environments and creating an ergonomic imaging apparatus (Pinz et al., 1998; Hill, 1999; Zana & Klein, 1999). Finally, retinal scanning is viewed as invasive by some individuals due to the light intrusion into the eye but also because personal medical conditions can be picked up (Moody, 2004).

The iris has been a source of investigation for identification for over a century, with Alphonse Bertillon in 1885 suggesting an iris colour classification system as a form of identification (1885). Although it may seem overly simplistic today, the approach was merely a logical extension of the classificatory approach to anthropometry evident in criminalistics at the time. More recently, much more detailed biological and anatomical work has been undertaken to allow the use of the iris in human identification and authentification contexts. Daughman converts the highly complex structure of the iris into a numerical string of numbers (1999, 2003, 2007). The iris, rather than being a simple flat ring of colour around the pupil, is actually composed of ligaments, arches, crypts, furrows and so forth (Daughman, 2007). These structural features are the result of the layered nature of the iris. Most posteriorly is a dark pigmented epithelium which is impenetrable to light, then towards the front of the eye one finds pupillary dilator and sphincter muscles, vascularised stroma and an anterior layer of chromataphores and melanocytes (Wildes, 1997; Daughman, 2007). It is at this layer that the varying levels of melanin determine the colour of the iris, with the absence of melanin resulting in blue eyes (Wildes, 1997; Daughman, 2007). As with fingerprints, the structure of the iris has some genetic influences, but is fundamentally formed in utero, and the unique intra-uterine environment ensures that the iris is therefore unique to each person (Daughman, 2007). Indeed even genetically related irises (that is, irises from the left and right eye) show the same level of difference as irises from completely unrelated sources (Wildes, 1997; Daughman, 2003, 2007; Ganeshan et al., 2006). The iris begins to form by the third month of gestation and is largely complete by the eighth month (Daughman, 2007). Although there may be slight variation until the adolescent years, iris structure shows good stability over time, varying only in old age with slight depigmentation and reduction in the pupillary opening (Wildes, 1997; Ganeshan et al., 2006).

The first challenge of this method of identification is to automate the location of the iris. Thankfully, the eye is relatively easy to locate within a face, but beyond this any automated software will need to locate the iris within the eye; it is very rarely in the centre of the eye, and can be offset nasally as much as 15 per cent (Wildes, 1997; Daughman, 2003). Furthermore, the location of the eyelids requires defining. This will remove some of the area of analysis, but only around fifty pixels across the iris are actually needed for analysis (Daughman, 2003). The relationship between the eye and the camera can also be influential (Daughman, 2003). The extremely rapid response of the iris to light can be used as a test of 'liveness' at biometric stations to reduce the potential of identity fraud with high resolution images of irises (Wildes, 1997). The comparison between any two irises is based on differences between the strings of numbers produced from each eye. This is not to say that this method does not have difficulties: the iris is small, the eye must be close enough to the camera to take a reading, the pupillary boundary can be difficult to detect and the distinct pattern of irises can be difficult to detect under normal light so near infrared must be utilised (Wildes, 1997; Daughman, 2003). A debate still takes place today regarding the overall stability of the iris with age and light (Marino et al., 2006).

Some of these issues can be offset by adopting different analytical approaches, such as laying a circular region over the area of the iris to focus analysis or utilising size relationships between structural features (such as borders) as in Ganeshan et al. (2006). One of the key reasons for so much attention in Daughman's algorithims is, however, their clear success in the field. His methodology has been deployed at border entry and exit points in the United Arab Emirates, which has resulted in more than 840,751 enrolments and 6.2 billion all-against-all daily comparisons (Daughman, 2007; Al-Raisi & Al-Khouri, 2008). The instigation of this system was the result of a particularly pressing problem in the United Arab Emirates of individuals ejected from the country re-entering using different documents (Al-Raisi and Al-Khouri, 2008). The system was fully implemented in January 2003 following a one-and-a-half year pilot test leading to twenty enrolment centres and twenty-seven border centres; 2.5 trillion comparisons have been made, each in under three seconds and boasting a zero false match rate in addition to a 100 per cent enrolment success rate (Al-Raisi and Al-Khouri, 2008). Despite the clear success, Al-Raisi and Al-Khouri note three problems with this system (2008). The first involved attempts to bypass the system with the use of eye drops to dilate the pupils. This was resolved by ensuring the system rejects individuals when the pupil to iris radii ratio exceeds 60 per cent. The second involved the lighting environment in the enrolment centres, which are now closely regulated. The third focussed on operator training rather than the technology itself. While iris scans are one of the most reliable forms of biometric identification, similarly to retinal scans they may be considered invasive by members of the public (Moody, 2004). Indeed, iris and retinal scans highlight some of the key concerns surrounding invasive biometrics today, and it is possible that retinal scans could also reveal personal information, such as medical conditions, that are not strictly pertinent to the identification context in question.

4.4   Body fat

The diet of an individual will affect all aspects of his or her life, but it is arguably most closely related to the activities and functioning of the internal organs. It is, after all, the function of the digestive system of structures and organs to distribute useful nutrients around the body to ensure optimum functioning. Scholars have long argued that socio-economic and lifestyle factors greatly influence diet. In the modern Western world, increased income results in better health and a reduction in fat, as represented by the Body Mass Index (BMI; weight in kg/height in m²). Wadden et al. highlight the five weight classes based on BMI values, with normal weight between 18.5 and 24.9 BMI, underweight less than 18, overweight between 25 and 29.9, obese between 30 and 39.9 and extreme obesity over 40 on the BMI scale (2006). BMI is a far from perfect measure premised on the notion of a 'normal' weight across all human groups (Halse, 2009). Akridge et al. argue that it is a poor index of fatness in children unless both age and sex are taken into consideration (2007). Hill et al. agree that BMI is not a perfect proxy for obesity, but argue that it can be improved with the addition of waist circumference measurements (2006).

Worldwide, body weight has been steadily increasing, with as many as 1.6 billion people over the age of sixteen now classed as medically overweight (Hebl et al., 2008). Obesity levels have risen sharply since the 1980s (Dietz, 2006; Averett et al., 2008), to the extent that the term 'obesity epidemic' is commonly employed (Hill et al., 2006). In Canada, childhood obesity levels have tripled to 26 per cent, while adult levels have increased to 23 per cent (Smoyer-Tomic et al., 2008). The highest increase in obesity levels occurs in the most disadvantaged groups (Kwate, 2008). Work has also shown that relatives and friends of obese individuals are more likely to become obese themselves (Dietz, 2006; Barry & Petry, 2008). It is important to appreciate that some increases and fluctuations in BMI are linked with development. BMI rises naturally between birth and nine months, drops to its lowest at 4 to 5 years and then continues to rise through adolescence (Dietz, 2006). During this last age range, much of the increase in BMI can be linked to the increase in quantity of non-fatty tissues (Akridge et al., 2007), while body fat itself tends to migrate from peripheral to more central sites (Dietz, 2006). There are some sex-related differences, but mainly during the rise in adolescence when female body fat increases from 17 to 25 per cent of overall body weight (Dietz, 2006). Fat distribution across the body also demonstrates sex differences, with females having fat deposited preferentially in the gluteal regions, and with males having fat deposited more centrally in the abdominal regions (Dietz, 2006). Excess body fat in the upper body is linked much more strongly to clinical complications and illness than that in the lower body (Wadden et al., 2006). Overall, surveys show that more men than women are overweight, but that more women than men are obese (Hill et al., 2006).

Higher body mass index has been linked with an increased risk of developing a number of pathological clinical conditions, including cardiovascular disease, diabetes and physical disabilities (Himes, 2000; Clark et al., 2007; Barry & Petry, 2008; Smoyer-Tomic et al., 2008), with mortality increasing over 25 kg/m^2 and then dramatically so over 30 kg/m^2 (Hill et al., 2006). Robertson et al. highlight the fact that of the seven major risk factors to mortality proposed by the World Health Organization, six relate to diet, with the leading factor directly related to obesity (2006). Ultimately, they argue that 'the greater the degree of obesity, the greater the risk of mortality' (Wadden et al., 2006: 1029), although it has been suggested that even moderately overweight individuals are at greater risk of physical and psychological illnesses (Barry & Petry, 2009). The relationship between obesity and health is, however, complicated by the correlation with social inequalities. Obesity tends to be higher amongst those of lower socio-economic status, and these individuals are already at risk of experiencing poor health outcomes through a variety of other aetiologies. Consequently, some of the figures relating to obesity and poor health may well reflect other class-related psychosocial stressors. Despite the considerable, well-publicised concerns over the current 'obesity epidemic', the relationship between BMI and health is far from straightforward (Halse, 2009).

4.4.1   Body fat and identity

In today's society, fatness is a 'social repugnant state that is a metonym for laziness and ugliness and an indicator of some troubling physical or psychological pathology warranting oversight, disciplining and correction' (Halse, 2009: 48). Halse (2009: 50–1) draws upon Rose and Novas' (2005) concept of the bio-citizen and discusses how in popular and medical discourse to be over-weight makes one a 'bad citizen' – similar to those who have an alcohol or sub-stance abuse problem. Overweight mothers are chastised for not only putting their own health at risk but those of their children (even unborn children). Fat bodies are conceptualised as pathological and those who are overweight are greedy, out of control and lacking personal discipline. To have a slim, toned and firm body is to project the 'right attitude' and suggests both control and care of oneself (Williams & Bendelow, 1998: 74). Responsibility for being overweight is placed on the individual – a similar feature to that noted by Crawshaw in other areas of health (2007, 2009). Despite the growing numbers of people classed as overweight, fat bodies continue to suffer stigmatisation, and obesity can have a profound effect on social identity (Murray, 2009).

Hebl et al. suggest that 'research has revealed that even more than a medical disease, being heavy seems to be a social disease' (2008: 46). Thus, research has been conducted into the prejudicial nature surrounding being overweight or obese. In examining perceptions of obesity across the life course, these authors suggest that overweight individuals are perceived as less attractive, less success-ful in personal relationships and less popular, while they are also selected into college, social groups and relationships less often and are viewed less favour-ably in employment situations (Hebl et al. 2008). Park et al. postulate an evolu-tionary undertone for these prejudices, suggesting that obesity can be viewed as a deviation from the species-typical morphological norm and may, therefore, be interpreted as a cue for pathogen infection (2007). Increasing age seems to reduce the severity of some of these negative perceptions (though perhaps replaced with the negative perceptions surrounding old age). Sex is also signifi-cant, with males attaching greater negative values to being overweight, espe-cially in women. Interestingly, overweight individuals attach the same degree of negativity to other overweight people and do not seem to demonstrate any more tolerance than thinner individuals.

Increased body mass has also been associated with psychosocial stressors, such as those associated with job issues, low self-esteem and poor quality of life (Barry & Petry, 2008). This also holds true during childhood, when neg-lect, abuse and ill treatment can increase the risk of obesity (Dietz, 2006). The resultant increase in BMI could be the consequence of increased consumption or decreased exercise, but it is also suggested that stress can directly affect body weight through activation of the hypothalamic-pituitary-adrenal axis. The degree of increase is argued to be very much dependent on the population demographic studied (Barry & Petry, 2008). In contrast, BMI can also drop due to anxiety and depression (Averett et al., 2008), perhaps due to the adoption of a less balanced diet. Eating disorders are essentially behavioural disorders that occur on a gradient of severity (Coughlin & Guarda, 2006). Females are more likely to binge eat and undertake strict diets or fasting protocols (de Souza Ferreira & da Veiga, 2008) and women are ten times more likely to develop anorexia nervosa (Coughlin & Guarda, 2006). The prevalence of eating disorders in women results from the increased vulnerability of women to the concept of equating beauty with slimness (de Souza Ferreira & da Veiga, 2008). Work suggests that eating disorders in males tend to focus on muscle building (vigorexia) rather than weight loss, using fat and protein-rich diets (de Souza Ferreira & da Veiga, 2008).

It has been found that being overweight can be associated with unemployment and that overweight individuals earn, on average, less than those with lower BMI values (Barry & Petry, 2008). Poverty also tends to be associated with isolation, boredom and depression, which can often result in snacking or sedentary behaviours (Darmon & Drewnowski, 2008). As discussed previously, there is a correlation in the developed world between low socio-economic status and obesity. 'The poor do not eat what they want, or what they know they should eat, but what they can afford' (PAHO, 2000), and the effect of poverty and wage income on access to nutritional resources should not be underestimated. In Europe, up to 10 per cent of the population officially lives in poverty (Robertson et al., 2006). Part of the issue is that the relationship between poverty and food intake results in a negative spiral. A lack of suitable nutrition impairs the ability to get good work and to concentrate fully in school, contributing to a reduced capacity for social mobility (Robertson et al., 2006). In terms of nutritional inputs, for example, it has been noted that higher socio-economic status is associated with higher intakes of whole grains, lean meats, fish, low-fat dairy products, fruit and vegetables, while lower socio-economic groups ate more refined cereals, fatty meats, added fat, salt, pasta, rice and potatoes (Robertson et al., 2006; Darmon & Drewnowski, 2008; Mpontshane et al., 2008). In some countries, the high cost of healthier foods has been negated with the introduction of food vouchers, and this has proven effective even in regions where dietary advice alone has had little impact (Darmon & Drewnowski, 2008).

Country of origin is highly significant, with lower socio-economic groups in some European countries consuming more fruit and vegetables (Darmon & Drewnowski, 2008). Also, tea, traditionally the drink of the poorer classes in the United Kingdom, has been shown to have positive benefits to overall health (Hamer et al., 2008). Studies show that a rapid transition from a traditionally lower socio-economic class diet of the Mediterranean style (high in fish, fruit and vegetables) to a higher socio-economic diet has brought with it an increase in cardiovascular disease in these regions (Robertson et al., 2006). In the developed world, there is an inverse relationship between socio-economic status and obesity, while the reverse seems true in the developing world. Regardless, even in countries where severe undernourishment is commonplace, obesity and low weight exist simultaneously and side by side (Robertson et al., 2006).

Other aspects of food-related identity include vegetarianism and veganism. The former is reported to be on the decrease (the United States saw a drop from 7 per cent to 2.5 per cent of the population from 1994 to 2000), though vegetarians still constitute a significant portion of the population (Johnston & Sabaté, 2006). In Western cultures, vegetarianism is usually associated with a reluctance to kill and eat animals, although its history can be traced back to Pythagoras and it is an integral component of a number of religions and faiths, such as Buddhism and Hinduism (Johnston & Sabaté, 2006).

Research has also linked alcohol consumption to BMI, and in women there is an inverse relationship between weight and alcohol use, while in men, although the results are arguably inconsistent, a positive relationship appears (Barry & Petry, 2009). Such work tends to produce a confusing image of the relationship between the two; for example, individuals with a low BMI often drink alcohol more frequently but in lower quantities, while women tend to substitute alcohol calories for other sources of energy while men do not, and various unhealthy activities are often linked together, blurring cause and effects (Hamer et al., 2008; Barry & Petry, 2009). Other addictive substances share complex relationships with body weight. Although drug use tends not to be associated with weight (which contradicts the commonly held belief that opiate use results in weight loss), the cessation of smoking does result in an increase in weight, especially in women (Barry & Petry, 2009), although this tends to be short term and after 10 years body size is similar to those who have never smoked (Himes, 2000).

Education has often been cited as the key limiting factor influencing diet choice. Lower levels of education are also associated with lower levels of interest in cooking (Darmon & Drewnowski, 2008). Furthermore, level of education has been related to method of child feeding which has, in turn, been linked to weight. Clark et al. argue that parents with a post-16 standard of education tend to use 'instrumental feeding' approaches more than those with compulsory education only, and tend to use fewer feeding strategies (2007). This tends to suggest that less educated parents may increase the risk of producing overweight children. Much work has shown, however, that restricting certain foods in childhood can result in dietary issues later in life, and work by Clark et al. shows that parents with higher education levels tend to restrict more food types than those with lower levels (2007). In addition, much of the published literature on diet and eating habits relies on self-reporting questionnaires.

Oftentimes these are very difficult to complete for those with low levels of education and thus may produce a form of sampling bias (de Souza Ferreira & da Veiga, 2008).

Age has proven a critical factor in terms of dietary intake. If, as Akridge et al. point out, obesity affects skeletal development in some way, then this will have an influence on the timing of any age-related clinical interventions, such as orthodontic treatments (2007). Obesity results in the early onset of puberty, accelerated growth and increased risk of breast cancer in girls and both early onset and delayed onset of puberty in boys (Akridge et al., 2007). Adolescence, in particular, has been shown to be a period of large potential impact of dietary habits. Indeed the number of overweight adolescents has increased by a multiple of three since the 1970s (Pollock et al., 2007). In their study of Brazilian adolescents, de Souza Ferreira and da Veiga note that the risk of obesity increases during this period (especially in lower socio-economic groups) as does the incidence of associated eating disorders (2008). Obesity in childhood has implications for adulthood with evidence suggesting that 25 per cent to 80 per cent of overweight children remain so as adults (Dietz, 2006). Calorie intake during development in utero can also be significant in determining overall adult weight (Ravelli et al., 1999), and breastfeeding infants are less likely to be obese later in life (Hill et al., 2006). Excess weight in the elderly can lead to increased damage to the joints, higher rates of osteoarthritis and lower aerobic capacity and muscle strength due to reduced activity (Himes, 2000). Increased functional limitations due to increased body weight seems to affect women more than men, and the greatest impacts focus on mobility such as walking and climbing stairs (Himes, 2000).

Researchers have long acknowledged a relationship between marriage status and body mass index with an increase in BMI following marriage (Averett et al., 2008). Cohabiting and recently divorced individuals often maintain lower BMI values (Averett et al., 2008). Much of these weight changes possibly link to the notion of the 'marriage market' and attracting a partner.

Ancestry also has an important role to play. Research in the United States has demonstrated that obesity rates are similar in white men and women, but that African American and Hispanic women have much higher rates than men (Hill et al., 2006; Kwate, 2008). Recent work on epigenetics is relevant here (see Chapter 6) in addition to variations in cultural practices. Socio-economic status does not seem to be the only factor here (Hill et al., 2006).

Access to supermarkets and other food stores is also significant in terms of diet and consumption. Historically, physical access to food stores was less of a limiting factor. In 1801, more than four out of five English people lived in rural areas, but by 1900 four out of five people lived in urban areas and relied on the transport of food to these locations (Burnett, 1976). Thus we moved from a point where food choice was determined by availability to one where choice was determined principally by cost (Burnett, 1976). One hundred years later, cost still has a significant impact on physical access to food and food suppliers. Work by Smoyer-Tomic et al. has shown that neighbourhoods of lower socio-economic status are closely correlated to the location of fast food outlets (2008). Further to this, work by Kwate has shown that there is a higher proportion of fast food outlets in predominantly black areas of US towns and cities (2008). This may, in part, be due to the lower rents, less competitive retail climate and less restrictive land use regulations in these areas (Smoyer-Tomic et al., 2008), but also to concentrated poverty, access to minimum-wage labour and less community political strength in these areas to resist the presence of fast food outlets (Kwate, 2008). In addition, the larger supermarkets which contain a wider range of food products tend to be in out-of-town locations, and lower socio-economic groups (including single parents) statistically have less access to cars or other means of transport (Darmon & Drewnowski, 2008; Smoyer-Tomic et al., 2008). Thus it may be more difficult to alter eating habits in these areas where cheaper, albeit less healthy, food is more readily available (Smoyer-Tomic et al., 2008). Time that can be dedicated to food shopping and preparation is also a factor (Darmon & Drewnowski, 2008; Smoyer-Tomic et al., 2008). Physical access to food products also links to illness and disease; food designed for diabetics, for example, is often not stocked in lower socio-economic neighbourhoods (Darmon & Drewnowski, 2008). It would be interesting to note the relationship between socio-economic group and the use of Internet food shopping, as this has the potential to bypass many of the physical restraints to access to healthy food.

It is important, however, to consider that not all increases in weight are due to environmental factors. Medical conditions, such as Cushing's syndrome, growth hormone deficiency, hypothyroidism, polycystic ovary disease and Turner's syndrome, can also result in an increase in weight (Dietz, 2006; Hill et al., 2006). Furthermore, medical psychiatric conditions like mood disorders, schizophrenia and the use of psychotropic medications can also influence body weight (Coughlin & Guarda, 2006).

Evidence of the impact of social environment and socially constructed identity on the health, well-being and form of the physical biological body is clear. Perhaps the clearest example of this is noted in Robertson et al.'s discussion of the political significance of diet in Europe (2006). Simply changing our lifestyle and our social circumstances is enough to alter our physical health, changing our social identity with regard to diet, which in turn results in a change in our biological identity and identification parameters.

4.4.2   Body fat and identification

The relationship between body fat and identity is also of significance in human identification contexts. This is particularly true of cases involving deceased individuals when the soft tissues are no longer preserved and facial reconstructions are required to aid identification. Techniques of facial reconstruction (see Wilkinson, 2004 for greater discussion of this) require that the soft tissue contours of the face are reliably reproduced from the underlying facial bones. This is carried out using soft tissue depth charts which have been assessed and established for numerous population groups. These charts, however, cannot take into account individual differences in the deposition of fat around the face, and it is not possible to determine from skeletal remains whether an individual was obese or slim. Fat can have a profound influence on facial appearance and has been a potentially confounding factor in reconstructions of this sort.

Within most human identification contexts the soft tissues decay relatively quickly and fat is no exception. In certain circumstances, however, fat undergoes a process known as saponification in which the fat deposits are converted into layers of adipocere. This layer of adipocere can be helpful, since it may protect and preserve the surface of the skin and features on it for longer than would otherwise have occurred. Although this process is well reported, the exact mechanisms of this change remain under investigation (Aufderheide, 2011; Ubelaker & Zarenko, 2011).

4.5   Organ transplantation

The successful transplantation of one human's organ into another human is often heralded as a miracle of modern medicine. In organ transplantation the body is not viewed holistically, but as a series of 'spare parts' that can be exchanged with others when they wear out (Hogle, 2003: 63). This mechanistic view has led to a conceptual fragmentation of the body in the view of medical practitioners (Brown & Webster, 2004). As several researchers have shown over recent years, this process has had considerable implications for the identity of the donor, the recipient and the donor's kin (Scheper-Hughes & Lock, 1987). The continuing life of a person's organ, even after death, raises questions concerning the bounded and finite nature of our bodies. Through the process of organ transplantation, an individual's bodily tissues come to reside in multiple spaces and by doing so they serve to create new identities while fracturing those previously held.

Examples of this are evident in the work of Sharp, who has studied in depth the processes and impact of organ donation on those involved from an anthropological perspective (2006). One of the most emotive and intriguing aspects of this research is the status of the donor. In order for transplants to succeed, the organs must be removed from a live body. Therefore, transplantation has only been possible since the development of life support equipment and artificial respirators. It has also only been practised since the creation of a new identity: that of the 'brain dead' patient. These individuals are those whose brains have been irretrievably damaged and who will not recover. They are usually the victims of car crashes, gunshot wounds and so forth. While their bodies are kept alive through artificial means, diagnostic tests indicate that there is no (or extremely limited) brain activity and the condition is irreversible. As Lock describes, 'Brain-dead patients remain betwixt and between, both alive and dead, breathing with technological assistance, but irreversibly unconscious' (2004: 136). The concept of 'brain dead' bodies, or 'living cadavers' as they were first known in the 1960s (Lock, 2004), was introduced specifically to allow the harvesting of organs. As such the concept of 'brain death' is essentially a culturally constructed one (Ohnuki-Tierney, 1994: 234). In describing the complexity of the phenomenon of brain death, Stern remarks: 'It must be understood not simply as a medical condition, but as a complex, cultural object, the meaning of which is necessarily unstable, since it is determined by so many different practices, ideas and beliefs' (2008: 350). Whilst this status is recognised in North America and Europe, in Japan it is not, and, prior to 1997, it was illegal to procure organs from 'brain dead' individuals in Japan (Lock, 2004: 136).

The definition of brain death in both medical and legal terms has been revised a number of times since its inception and is now widely, though not universally, accepted (Stern, 2008: 349). As Sharp observes: 'Donors are particularly liminal beings caught somewhere between patient and cadaver status' (2006: 4). This uncertain status becomes manifest in the difficulties the medical profession has faced in settling on unproblematic terminology for these patients: while it is difficult to identify them as corpses, other labels, such as 'beating-heart cadaver', are not viewed as acceptable either (Ohnuki-Tierney, 1994: 234). The ambiguities surrounding this new identity are exemplified by the anaesthetising of brain dead bodies prior to organ removal. There is some debate within the medical community about this practice and why, if the person is dead, it is necessary at all. Some practitioners argue that residual spinal cord activity can cause the body to move during the process of organ removal and this would be upsetting to operating room staff and cause doubts as to the brain dead status of the person (Sharp, 2006). The harvesting of organs from a live, albeit brain dead, body is an intense experience for those involved (see Lock, 2004), and it could be that the anaesthetic is as much about enabling medical staff to deal emotionally with this somewhat Frankensteinian and grotesque act. Others have justified the use of an anaesthetic by invoking the concept of death as a 'process' rather than 'a moment in time' and one needs to ensure that there is no possibility of the donor feeling any pain (Sharp, 2006). This latter explanation also reveals that despite medical advances (and ironically during the midst of one of the most renowned manifestations of this – organ transplantation) the human body still holds some mystery and the application of anaesthesia covers for the remote possibility of these unknowns (Sharp, 2006).

The diagnostic parameters of establishing the brain dead status of a patient has been disputed since the inception of the term (Angrosino, 1994; Lock, 2004). The medical community does not have clearly defined, universal standards for diagnosing brain death. The biology of death is subject to cultural interpretation. Angrosino argues that 'the decision to declare brain death is a social act that occurs in various ways depending on shifting circumstances, it is not purely an exercise of neutral biomedical technology' (1994: 243). While patients may be deemed irreversibly brain dead, they are not biologically dead; their hair and nails may continue to grow and some may still show clusters of brain activity. Infants are born to brain dead mothers. Indeed, if these individuals' organs are to be used for transplantation then they must be prevented from dying (Lock, 2002, 2004). It is particularly difficult for relatives to reconcile a diagnosis of death when faced with the warm, breathing body of their loved one (Stern, 2008). In her interviews with thirty-two intensive care unit specialists, Lock found more ambivalence about the status of the brain dead individual than the specialists initially admitted (2004). As Ohnuki-Tierney observes, the question of when life ends is complex and culturally defined and the concept of 'brain death' has brought with it a 'different sort of death', one that fails to mesh with what many perceive as 'the end' (1994: 235; see also Kaufman & Morgan, 2005).

The cadaveric human body has become a highly lucrative entity: as many as 150 parts can be reused, and an individual body alone is worth more than $230,000 on the open market (Sharp, 2006). The commodification of body parts has, inevitably, led to some dubious, unethical and even illegal practices regarding their procurement and transplantation. The global and illicit trade in human organs has been highlighted by Scheper-Hughes (2000, 2011) and is viewed by some as another means of exploiting the poor and disadvantaged. What of the organs themselves? As Scheper-Hughes states: 'While for transplant specialists an organ is just a "thing", a commodity better used than wasted, to a great many people an organ is something else – a lively, animate, and spiritualised part of the self which most would still like to take with them when they die' (2000: 224).

Sharp's study of organ transplant recipients made clear that for many, the transplanted organs retained the identity of the donor within the recipient (2006). The organs 'retain a substantive connection with their source after separation' (Lambert & McDonald, 2009: 3). Organ recipients will sometimes undergo changes in routine and behaviour which align with the routines and behaviours of the donors (Lock, 2002; Brown & Webster, 2004). Recipients feel a sense of relatedness with the donors and their families (Sharp, 2006). Human organ transplantation is a highly emotive and much debated subject, particularly if the donors must be of 'living cadaver' status. It feeds into the often asked question as to why people are unwilling to part with their organs upon death. The answer likely lies with the notion that such a practice tests fundamental certainties that we hold dear, including the moment of death and the boundedness of our being.

The transformation of the body into a commodity is not a modern phenomenon, but also occurred in eighteenth-century Britain as a consequence of the increasing importance of dissection in medical research. As Cherryson discusses, bodies used for dissection were often those of the unclaimed poor or criminals, but high demand from the medical establishment provided a lucrative business for grave robbers. At this time, the soul of the dead person was thought to linger for a few days after death and dissection led to an unease about the impact of this ultimate fragmentation of the body by anatomists on resurrection and the afterlife (2010).

Examples of similar concerns about bodily integrity and 'wholeness' are illustrated in the burial of individuals with limbs that had previously been amputated during life (Chapman 2010). By contrast, the deliberate fragmentation of the body as part of a funerary process (e.g. through cremation or excarnation) and the distribution or intermingling of these fragments with the body parts of others is a feature of the funerary rituals of many societies, both past and present. John Chapman discusses the distribution of body parts in Neolithic Europe as a means of materialising kinship links (2010). Parallels can be drawn between the relationships and identities constructed through the fragmented and exchanged body that appears in discourses on organ transplantation and the relationships of 'enchainment' suggested for the Neolithic.

4.6   Conclusion

On the face of it, the internal organs seem to have less to do with our identities and identification than they do with the effective functioning of our bodies, and as a consequence collectively they receive less attention than, for example, the skin or the skeleton. This omission, however, drastically simplifies the situation. As we have seen, our soft tissues and internal organs are involved in a range of approaches and techniques of identification, from those popular historically (such as blood typing) to those at the forefront of current biometric approaches (such as vascular patterning and iris scanning). Beyond their growing use in identification, the internal organs, the way in which they are encased within our bodies and within our physical selves, have implications for a number of aspects of our identities. Blood, for example, has great significance for those with differing religious identities, to the extent that it may determine which potentially life-saving clinical treatments can be undertaken (Jarvis & Northcott, 1987). Because our health is so profoundly influenced by our identities, our organs embody these identities in a variety of pathologies that relate to our socially mediated activities, diet and living environment. Body fat too is significant, and here we have seen how the presence of fat throughout the body influences many aspects of identity. Body fat is not fixed; one can gain or lose weight, producing dramatic changes to one's outward appearance, and thus it features strongly in discourses of identity and control. With this particular bodily component, it is surprisingly clear how the biological and social worlds of an individual coalesce and interact, not only within the self, but also from the perspective of others. Perhaps their greatest interest lies with the first sentence of this conclusion – that they are so closely associated with the functioning of the body. Thus, perhaps with the internal organs more than any other part of the body, they tightly bind who we are and how we determine who we are with the functioning of our physical self and with the life-sustaining components of our bodies. This cannot be any clearer than when we examine the social repercussions of organ transplantation – the provision of life-sustaining organs from one person to another. In the following chapter, we will see that a number of these themes continue as we examine the biological structure that our soft tissues surround, our skeleton.

5

The skeleton

The bony frame of the human body provides structure, support and protection for the soft tissues and the fulcra to facilitate movement. Bone is a living tissue, and as such is supplied with oxygen and nutrients necessary for its effective functioning. As well as structural support, bones and the marrow within are responsible for providing the body with red and white blood cells, and also serve as an energy and mineral store. The skeleton, once fully grown, has long been erroneously conceptualised as a rigid, fixed and unchanging structure. In actuality, the skeleton is a dynamic plastic tissue, able to grow and respond to the physiological requirements of the individual and external stimuli throughout the life course (Agarwal & Beauchesne, 2011). The human skeleton has long been of interest to researchers working in anatomy, medicine, anthropology and archaeology. In terms of human identification, the skeleton is responsible for essential biometrics, including our height and facial architecture. Today, radiographs of the hands and teeth are commonly used to estimate a minimum age for living individuals in cases requiring the establishment of criminal liability (e.g. Schmeling et al., 2000), and dental records are used for positive identification in a range of scenarios. The vast majority of human identification work on the skeleton relates to deceased individuals, usually those whose bodies have partially or completely decomposed. Within bioarchaeology, the skeleton has long been the primary unit of analysis, as soft tissues survive only very rarely over long time periods. The importance of the skeleton as a repository for social information, however, has been a particular focus of recent archaeological research. Within the discipline of archaeology, funerary evidence provides a rich arena for the interpretation of both the biological and the cultural remnants of past populations. In this context, skeletal evidence for health, diet and mobility can be examined in conjunction with cultural variables (e.g. grave goods) from the funerary context in order to make inferences concerning aspects of social identity (e.g. gender, age, ethnicity). A number of studies have highlighted the dynamic relationship between the human skeleton and individual and group identity (e.g. see Gowland & Knüsel, 2006; Sofaer, 2006; Knudson & Stojanowski, 2008, 2009; Agarwal & Glencross, 2011); new methodologies to study the intricacies of the skeleton are now combining with a more thorough integration of social theory. The interrelationship between the skeleton and social identity is, however, still relatively under-explored within the medical and forensic sciences (though see for example Thompson, 2001; Crossland, 2002, 2009a, 2009b). This chapter describes the physiology of the skeleton and discusses the importance of these bony structures for the interpretation of human identity and identification.

5.1   The structure of the skeleton

Bone is a composite material, with around one third of its dry weight deriving from collagen. The remaining two thirds are formed of inorganic material, particularly hydroxyapatite. Apatites are highly adaptable chemical forms, and many varieties exist. Those found within the body are referred to as bioapatites; the proportion of the organic component is higher in juvenile skeletons (Baker et al., 2005). One cannot approach the skeleton as if it were a single structure. The adult skeleton comprises 206 bones, though the structure as a whole tends to be divided into different regions in order to assist analysis. Traditional divisions include separation of the cranial and post-cranial regions (the skull and everything suspended below the skull) or the axial and appendicular skeleton (the skull, spine and girdles and those skeletal elements suspended from this core axis). Macroscopically, the bones themselves can be assigned into different categories based on their shape and main functions. These categories include long bones, short bones, flat bones, irregular bones and sesamoid bones (detailed descriptions can be found in a number of textbooks including White & Folkens, 2005). Table 5.1 briefly summarises the key differences and provides examples. When we move towards the microscopic scale, we can see different forms of bone throughout the skeleton, the distribution of which is again determined by functional pressures across the body (see Figure 5.1).

Many texts describe the structure and development of bone, and useful examples include Baker et al. (2005), Scheuer and Black (2000) and McGowan (1999). The following descriptions are summaries of these works. Bone development essentially occurs in one of two ways, either by endochondral or intramembranous ossification. That is, either by converting a cartilage template into bone, or by laying down bone within a membrane outline. Occasionally, bones can develop through a combination of these approaches. The first bone material

Table 5.1 *Categories of bone based on form and function*

Bone type	Characteristic shape	Key functions	Examples
Long	Long axis with flared ends and medullary cavity	Movement, support	Femur, tibia, humerus
Short	Small with a high proportion of cancellous bone	Support	Carpals, tarsals
Flat	Flat and thin	Muscle attachment, protection	Scapula, sternum
Irregular	Varies, but complex with aspects of the other categories	Varies, but support, protection and movement	Vertebra, sphenoid
Sesamoid	Rounded and found within tendons	Facilitate movement while protecting joint	Patella

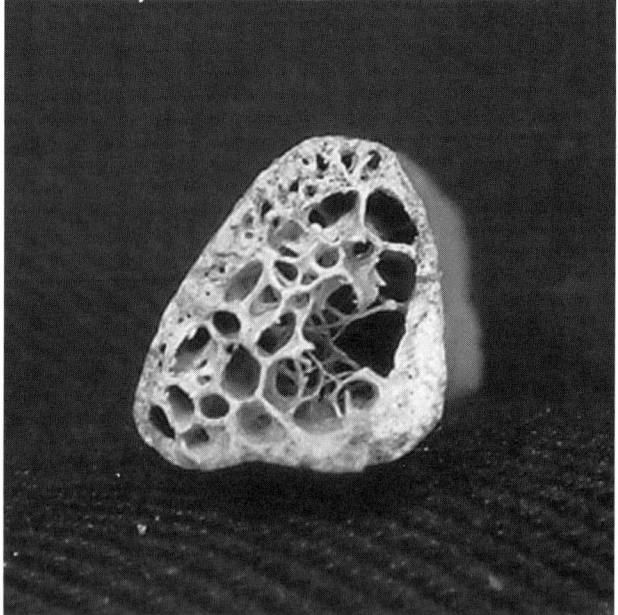

Figure 5.1 Cross-section through a rib showing the cancellous (spongy) bone providing support inside and the compact bone around the external surface. Courtesy of Anwen Caffell.

produced is referred to as woven bone, which has a disorganised structure and is mechanically weak. Over time, this woven bone is replaced by stronger, organised lamellar bone. The compact and cancellous bone described later in this chapter are the two forms of lamellar bone. Bone forms at sites known as the primary ossification centres, and generally these appear in utero. Later, many bones develop secondary centres of ossification and examination of appearance and fusion of these is an extremely useful age-at-death estimation technique. At birth, Baker et al. state, 450 primary ossification centres exist, which gradually fuse into the 206 bones of the adult skeleton (2005).

Compact bone is the dense, rigid bone that surrounds all bony elements. It provides strength and support and is thus seen in higher quantities in those bones that carry the greatest strain, such as the long bones of the lower limbs. Contrary to appearances, compact bone is not solid but is perforated with small channels that form what are known as Haversian systems. Each Haversian system, or osteon, is composed of a central channel surrounded by rings of bone called lamellae. Running perpendicular to this central channel and penetrating the bone are smaller tubes called Volkmann's canals. Lacunae, essentially small cavities, are found within the lamellae and have small canaliculi radiating from them. Collectively these effectively link Haversian systems together into one complex network. This network of channels and tubes allows for blood and nutrients to move through this otherwise dense bone. Both primary and secondary bone can be found with this structure. Primary bone is laid down during early development whereas secondary bone is deposited later and overlays and punctuates the primary bone.

Spongy or cancellous bone is considerably less dense than compact bone. It consists of a fine scaffolding of bone trabeculae (see Figure 5.1). Although individually weak, the combination of trabeculae works to provide tensile strength, responsive to multidirectional strain. Thus it is common within bony elements at the sites of joint or movement. The porous nature of cancellous bone means that nutrients can be absorbed from the bone marrow that surrounds it, and as such no Haversian systems are required.

Development of bone is referred to as modelling and relies on a balance between bone formation and resorption. Three key cells influence the creation and remodelling of bone. Osteoblasts are the bone-forming cells and lay down osteoid matrix during life. Once finished with their bone formation duties, a proportion of these osteoblasts will convert to osteocytes. These cells are found in the lacunae and serve to maintain nutrient flow. The third key cell type consists of osteoclasts, which are the bone-destroying cells. Although both bone cell types have an important interrelated role to play in bone modelling, their origins are quite different. Thus, bone modelling is the result of osteoblasts and osteoclasts (Frost, 2004). Osteoclasts derive from hematopoietic stem cells (like macrophages), whereas osteoblasts derive from mesenchymal stem cells (such as with fibroblasts and chondrocytes) (Matsuo & Irie, 2008).

The exact pattern of formation-resorption varies depending on bone type (long, short, irregular, etc; see Table 5.1). Endochondral ossification, which is common in long bones, begins in the bone shaft at an area with high vascularity within the cartilage template. From there, new woven bone is laid down within the mid-shaft, on top of which is created the periosteum. This ultimately surrounds the entire bone and is a tough fibrous membrane that functions as another bone-producing site. Cartilaginous growth plates appear at the ends of the bone shaft and it is here, where these cartilage cells are converted to bone cells, that longitudinal growth occurs. These growth plates separate the shaft of the skeletal element from the epiphyses, or secondary ossification centres. Bone deposition and resorption continues to occur on both the periosteal and endosteal bone surfaces, allowing for a thickening of the bone. Eventually the cartilage cells in the growth plate will be completely converted to bone and the epiphyses will have fused to the main shaft. Significant longitudinal growth can now no longer occur at this fused site. Intramembranous ossification, on the other hand, is a relatively straightforward affair whereby the membrane template is converted to bone directly and is common in many of the flat bones of the cranium. New bone can then be continuously deposited on the surface, thereby allowing expansion.

The influence of mechanical loading on overall bone structure was noted as early as 1892 by Julius Wolff (cited in Frost, 2004), but it is important to appreciate that load-bearing bones are not limited to weight-bearing bones and it is thought that it is these loads, rather than weight per se, that most influence bone strength after birth (Frost, 2004). These loads generally originate from muscle action (Frost, 2004), such as those complex muscle actions associated with walking (Kaptoge et al., 2007). Bone modelling occurs at two scales, either at the whole bone level or at specific locations along the bone. At the whole bone level, the actions of bone cells result in what is termed 'bone drift' during which the entire shape of the bone migrates to accommodate stresses and strains (Frost, 2004). In the femur, this has been shown to be further influenced by sex, age and activity (Kaptoge et al., 2007). Specific re/modelling occurs as a response to microscopic fatigue damage which is then repaired (Frost, 2004). It is thought that small strains at the point of the healing callus help to initiate bone cell activity, although the actual means through which this occurs is still unclear (Frost, 2004). McGowan also argues that the presence of osteons has an important role to play in halting small cracks and fractures from spreading across the bone by effectively introducing discontinuities into the material (1999). It is still fair to say that workers do not fully understand the mechanism of this modelling activity.

5.2   Skeletal aspects of identity and identification

Within forensic and archaeological contexts, human bone specialists create what is referred to as an osteological profile from preserved skeletal remains and/or radiographs. This profile comprises a description of the key bio-logical features of an individual, including sex, age-at-death, stature, ancestry and evidence of pathology or trauma. The ability to establish an individual's identity from his or her skeletal remains is dependent on preservation and com-pleteness. Within archaeological contexts, the osteological profiles of multiple individuals from cemetery samples are aggregated and analysed with the aim of addressing broader population-based questions of demography, living envir-onment, diet and lifeways of past peoples. Within a forensic context, interest is very much focussed on the individual. A frustrating phenomenon in this con-text is that, following the submission of this profile to the police, some years may pass before positive identification occurs, if at all (Steadman & Haglund, 2005). There is a range of different techniques for producing an 'osteoprofile', and the past decade has seen an increasing emphasis on the standardisation of these techniques in order to improve the comparability of osteological stud-ies. Examples of these methodological standards include those by Buikstra and Ubelaker (1994) and Brickley and McKinley (2004). There has been considerable work in the forensic field focussing on the same goal with discussions occur-ring in Europe and the United States (see summaries and rationale in the likes of Skinner et al., 2003; Cattaneo, 2007; Grivas & Komar, 2008; and Christensen & Crowder, 2009).

5.2.1   Biological sex

In an analysis of the human skeleton, sex is the first biological char-acteristic determined. This is because methods of estimating other character-istics, such as age-at-death and stature, are sex dependent. In addition, in the forensic field, knowing the sex of a deceased individual immediately rules out a large proportion of possible identifications. Sex is determined through an examination of the sexually dimorphic features of skeletal size and shape. The pelvis is the most sexually dimorphic element due to obvious functional differ-ences and therefore the most useful component of the skeleton for assessing sex (Buikstra & Ubelaker, 1994; Walker, 2005). The female pelvis is generally broader, with a larger pelvic inlet to facilitate pregnancy and childbirth. Since this is the only region of the skeleton where there is such a functional diffe-rence, the pelvis proves the most important bone for determining biological sex. The accuracy of sex determinations from the skeleton based on the pelvis alone is usually placed at around 96 per cent (Mays & Cox, 2000). Accuracy is variable and is dependent largely upon the degree of preservation or fragmen-tation of the pelvis and also the level and form of sexual dimorphism within a particular skeletal population, which can be highly variable (Walker, 1995).

The shape of the female pelvis has often been described as a compromise between optimum biomechanical functionality for bipedalism and the human requirement to give birth to relatively large-headed infants (see Chapters 1 and 2 for a discussion of the appropriateness of this concept). Hogervorst et al. state that the pelvis is 'the most defining skeletal element to read human evolution', arguing that childbirth was a stronger driver for evolution of the pelvis than the adoption of bipedalism (2009: 1). The changes in the pelvis over evolutionary time scales are significant in terms of the physical body and subsequent functioning (Stone et al., 2007). In particular, Hogervorst et al. draw special attention to the development of features that have come to be so important in sex estimation techniques (2009). As with many other aspects of functional anatomy, the male pelvis has been conceptualised as the norm; the ideal standard from which the female pelvis must deviate in order to fulfil reproductive functions (Schiebinger, 1986; Stone & Walrath, 2006).

The skull is also considered useful in accurately determining sex, with general estimates indicating an approximately 80 per cent success rate based on this alone (Mays & Cox, 2000). As with the pelvis, the accuracy of sex estimation is dependent on the state of preservation as well as the degree and range of sexual dimorphism (see Figure 5.2). A large sample of skeletons from a given population enables the osteologist to broadly establish these parameters (Walker, 2008). Such comparisons are usually possible within an archaeological context where investigators often work with cemetery samples, but within a forensic context this may not be the case. The features of the skulls are characterised as masculine or feminine, for the most part, according to degrees of robusticity or gracility. Overlap occurs between the sexes, and it is not uncommon for individuals to exhibit a mosaic of characteristically 'masculine' and 'feminine' features. As with the pelvis, most often sex determinations are based on subjective visual assessments of the features of the skull. Consequently, inter-observer discrepancies are common (Walrath et al., 2004). Osteometric techniques based on statistical analysis are being used more frequently, but the vast majority of anthropologists continue to use visual assessment (Walker, 2008).

Generally, morphological differences elsewhere in the skeleton result from the fact that males have a longer period of growth than females and therefore tend to have larger muscles and hence larger muscle attachment sites on their bones. The bones of males are often more robust than those of females, although there is overlap between the sexes. Overall size differences can also be measured to examine sex differences. The longer period of growth in men stems from the fact that puberty begins and ends 1 to 2 years later (Scheuer & Black, 2000). Most of the skeleton has been examined and tested for its ability to differentiate reliably between the sexes and such studies have met with varying success. Ultimately the pelvis is the most accurate feature, as differences

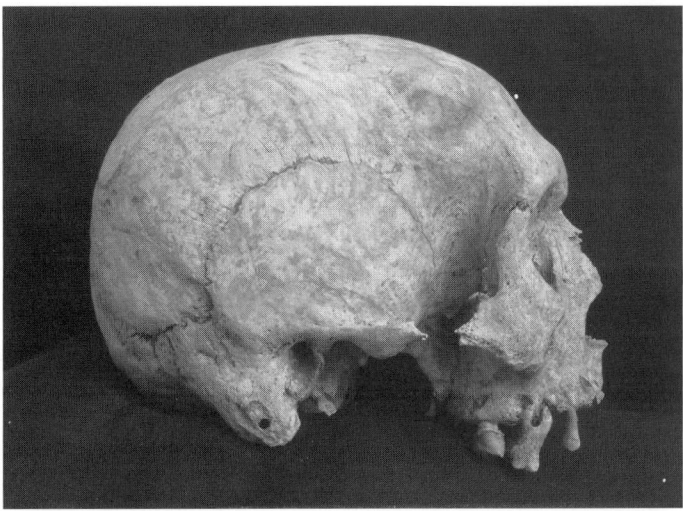

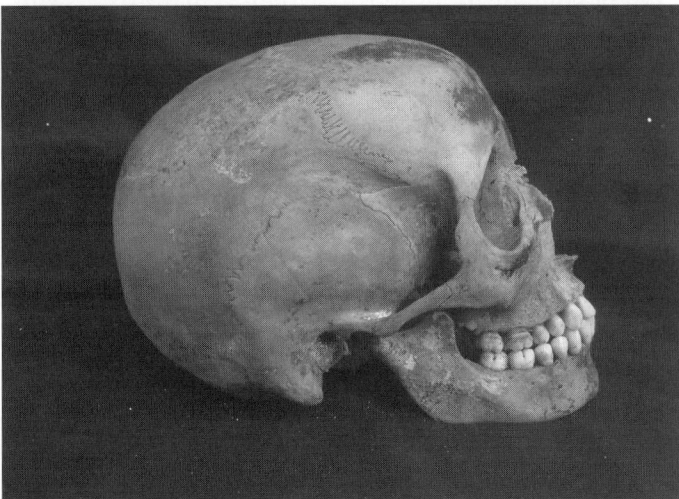

Figure 5.2 Archaeological skulls displaying typically male (top) and female (bottom) morphological features. Courtesy of Anwen Caffell.

here are morphological in character rather than reliant on degree of robusticity. Sex estimation is greatly enhanced when the entire skeleton is available for analysis. It is important to note that all of the bony features used to estimate sex fall along a continuum from male to female rather than in two discrete categories (see later in this chapter for more discussion).

A number of studies have attempted to estimate reliably the biological sex of children from a variety of skeletal indicators, most notably the morphology of the ilium, mandible and teeth (e.g. Weaver, 1980; Schutkowski, 1993;

Loth & Henneberg, 2001; Vlak et al., 2008; Cardoso, 2010; Wilson et al., 2011; and see Lewis, 2007). Results, however, vary depending on the population examined, and blind tests of these studies on different, known sex samples have shown that they are insufficiently reliable. For example, much work has utilised the valuable skeletal collection in Lisbon, where age, sex and other background details of the deceased are known. Even using this resource, the success of sexing juvenile skeletal remains is too low to be of practical use in real-world contexts (Vlak et al., 2008; Cardoso, 2010; Gonçalves et al., 2011; Wilson et al., 2011). It may well be that growth has more of an influence on morphology than sex does (Lewis & Rutty, 2003). Sexual dimorphism in the skeleton does not become marked until post-puberty (Baker et al., 2005), and difficulties may still arise when attempting to sex, for example, young adult males, whose skeletons are often less robust than older males and are thus more likely to be sexed as female (Walker, 1995, 2005). Indeed, once the skeleton reaches physical maturity it does not then remain fixed. The skeleton is fluid and constantly remodelling and key features of the pelvis and skull used to estimate sex have been shown to alter throughout the adult period and into old age, confounding identifications and resulting profiles (Walker, 1995, 2005).

As with all aspects of human identification, the goal when creating a skeletal profile is to record and categorise the human body. As always, the human body is not as readily categorised as we may wish or as the results of such analyses would appear to imply. Sex should be conceptualised as existing along a continuum from hyperfeminine to hypermasculine, rather than as discrete groups of male and female. For sex estimation there is some recognition of this continuum in terms of the sexing categories employed by osteologists, which include five groups: female, probable female, sex unknown, probable male, male. The use of categories in addition to 'male' and 'female' allows for the fact that it is not always possible to definitively assign a sex from skeletal remains. Further, the examination of a number of these sexually dimorphic features from around the body allows one to cope with fragmented, altered or incomplete remains. Environmental factors, however, appear to have a strong influence on the expression of sexually dimorphic skeletal traits, to the extent that they can vary substantially between different populations and even within populations over relatively short spans of time (Walker, 2008). For example, Walker discusses the lack of dimorphism in the mastoid process (a particularly useful skeletal feature for estimating sex) amongst some Californian Indian groups in a study which also illustrated differences in sexual dimorphism between these and European/American groups (see Figure 5.3) (2008). The variation in sexual dimorphism between spatially

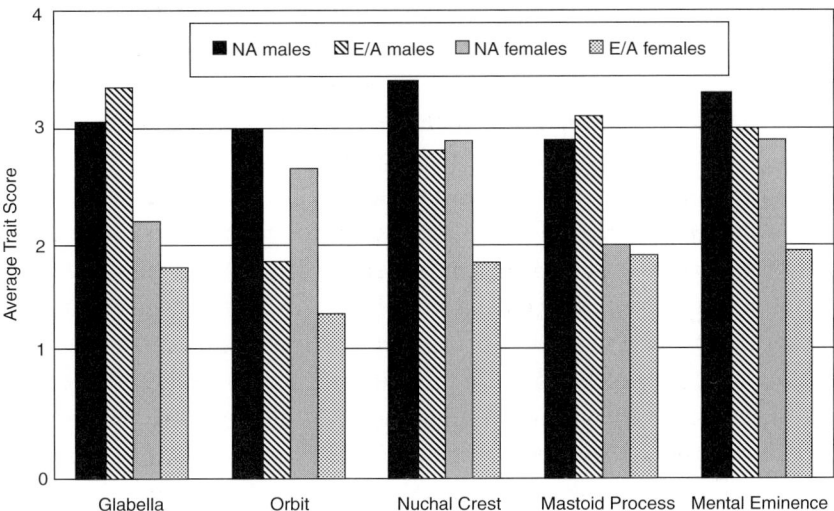

Figure 5.3 Average cranial trait scores of males and females for a modern European/American skeletal sample and an archaeological Native American skeletal sample. Cranial traits were 'scored' on a scale of 1 to 5, from gracile to robust (data adapted from Walker 2008, Fig 2., pg 48).

and temporally distant groups can be extremely marked and may result in a substantial and largely unquantifiable source of error when creating an osteo-profile. This skeletal variability is also apparent in other aspects of skeletal analysis. As discussed previously, skulls often show a mosaic of 'masculine' and 'feminine' skeletal features, thus potentially leading to ambiguity and difficulties in sex assignation; age-related alterations in the morphology of the pelvis and skull can further complicate sex estimation in younger and older adults (Walker, 1995, 2005). This does not reflect methodological failure for, as Epstein noted (see Chapter 2), the same is true of other physiological criteria: '*sex differences, like all differences in nature, lie on a continuum, and they become evident through statistical aggregation: there is no unambiguous dividing line between the two sexes*' (2004: 195).

While many skeletal features generally do aggregate statistically into two male/female clusters within any given population, a significant overlap exists between the poles, which can cause uncertainty in sex estimation (see Geller, 2008). As Sofaer discusses, this is not to say that anthropologists are unsuc-cessful at identifying sex, or that skeletal differences between the sexes are an 'irrelevant mirage', but that there is more biological overlap than is often acknowledged (2006: 96). In another example of biological ambiguity, the fact that an individual of either sex may show a mosaic of features indicates that the characterisation of such features in terms of biological sex alone is a somewhat flawed scientific construct; this will be discussed further in relation to gender.

5.2.2   *Sex, identity and the skeleton*

The relationship between biological sex as ascertained from skeletal remains and gender is complex and culturally situated. Interpretations of the sexed and gendered skeleton must also consider the social dimensions of age, ethnicity, status and even religion (Hollimon, 2011). As discussed in Chapters 1 and 2, it has been argued that only since the eighteenth century have anatomists begun a formal demarcation of bodily sex differences (Laqueur, 1990; Geller, 2008). Prior to this time, males and females were viewed as anatomically similar (Laqueur, 1990; though see Stolberg, 2003 for a critique). This is not to say that anatomists prior to this time did not recognise differences between males and females, but rather that the sexes were not conceptualised as opposites as they are today (Sheldon, 2002). The asexual conception of the skeleton prior to the eighteenth century is illustrated by the work of renowned German physical anthropologist Blumenbach, who famously coined the term 'Caucasian'. Blumenbach classified human phenotypic variation into five racial types. Three of the 'type specimens' for these racial groups were based on the skulls of females; when they were presented as fully fleshed portraits, however, they were all depicted as males (Schiebinger, 2004: 152–4).

During the eighteenth century, anatomists started for the first time to draw specifically female skeletons rather than simply a generic skeleton (Schiebinger, 1986). Representations of female skeletons began to be styled with exaggeratedly small skulls and large pelves. In such a way they were conceived as deviations from the idealised masculine norm and the male body became conceived as the prototype of the human body (Schiebinger, 1986; Tarvis, 1992; see also Stone & Walrath, 2006). This characterisation of the female skeleton aligned with the world view of 'woman' at that time as both intellectually inferior and defined in terms of their reproductive role (Schiebinger, 1986; see also Jordanova, 1989). This intellectual inferiority, as inferred from women's smaller skulls, also led to an association between females and so-called primitive or lower races (who were likewise perceived from cranial studies as inferior to 'whites'). In so doing 'women became racialised and lower races feminised' (Ahmed, 2002: 46). Later scientists recognised that cranial size was relative to body size and thus adjusted their calculations accordingly. It was then discovered, however, that contrary to ideology, women often had relatively larger cranial capacities than men. This led to the 'airhead' hypothesis, whereby some of this capacity should be accounted for by women having large air spaces within their skulls (see Gould, 1997)!

The skeleton is never purely biological in the sense that we imbue it with culturally situated meaning. Anatomists and anthropologists working in the nineteenth century helped to construct a biological reality of the female sex through their studies of cranial capacity. Gould's *Mismeasure of Man* illustrates the power of social and political context in influencing supposedly objective scientific analysis (1997). In relation to human identification, the relevance of social context for the interpretation of skeletal evidence, in particular sex estimation, has been discussed by a number of authors in recent years (e.g. Geller, 2005, 2008; Sofaer, 2006; Hollimon, 2011). Critiques have suggested that the scientific terminology currently used to describe male and female skeletal features feeds contemporary views about masculinity and femininity. For example, features of the male skull and pelvis are described as 'rugged', 'robust' and 'heavy', while those of females are described as 'delicate', 'gracile' and 'fine'. This use of language in the construction of difference between male and female anatomy has been noted by Peterson in his study of editions of *Gray's Anatomy* from the mid-nineteenth century to the present day (1998). Schiebinger (2004), Laqueur (1990), Oudshoorn (1994) and Spannier (1995) have also described this phenomenon within other scientific discourses. This use of language creates a polarisation of skeletal sexual characteristics which belies the considerable overlap between them. In an early study of sex estimation of skeletal remains, Weiss argued that skulls that exhibit 'intermediate' features were more likely to be sexed as males than female; thus contemporary perceptions associating femininity with gracility permeate supposedly objective scientific techniques (1972). As discussed previously, however, this is not to say that we cannot sex skeletons or that there are not real empirical differences between them. Repeated blind tests of the accuracy of sexing on skeletal remains have demonstrated that this can be achieved reliably for the majority of skeletons examined (e.g. Molleson & Cox, 1993).

In archaeological and forensic analyses the skeleton is usually conceptualised in terms of biological sex rather than gender. The skeleton is never fully 'biological' in terms of sex due to the way in which society constructs gendered difference and the impact that this has on bone physiology.

> One cannot easily separate bone biology from the experience of individuals growing, living and dying in particular cultures and historical periods and under different regimens of social gender.
> (Fausto-Sterling, 2005: 1510)

The role of gendered cultural practices for contributing to the degree and nature of sexual dimorphism within any one population has not been examined in any detail to date, though numerous studies have observed strong inter-population differences that cannot be explained simply in genetic terms. However, some authors have described ways in which the skeleton becomes 'gendered' in terms of skeletal pathologies and imprinted by the culturally con-structed roles ascribed to different sexes. For example, within societies male and female infants may receive different treatment in terms of care and diet from birth onwards due to perceived cultural value, thus exposing infants to differential risks of nutritional deficiencies and infectious diseases that may impact on growth, stature, morbidity and mortality in later life. Merbs was one of the first bioarchaeologists to describe variations in patterns of joint disease between males and females in terms of a gendered division of labour (though the word 'gender' was not used in archaeology at this time) (1983). Sofaer (2000, 2006) and Geller (2008) describe the way in which the skeleton can be likened to material culture in that it is shaped and moulded by cultural practices. A number of authors have examined pathological evidence on the skeletons from past populations with the aim of elucidating past gender roles. Sofaer's study of spinal degeneration in skeletal remains from the Isle of Ensay, in conjunc-tion with historical information, demonstrated that females exhibited greater frequencies of degenerative joint disease as a consequence of their gendered task of carrying heavy loads on their heads (2000). Perry's study of musculo-skeletal stress markers in a pre-Hispanic south-western population identified male and female patterns which could then be related to particular gendered activities (2004). Interestingly, Perry also noted males showing female patterns, possibly indicating gendered identities beyond the traditional binary dominant in Western discourse (2004). The identification of 'third genders' from archaeo-logical mortuary evidence through the burial of male skeletons with female grave goods and vice versa, or due to non-gender-normative patterns of skeletal pathologies, have been discussed by a number of authors (e.g. Hollimon, 1997; Knüsel, 2002). Other studies comparing bioarchaeological evidence between the sexes include analyses of differences in fracture patterns, infectious dis-eases, dental wear, dental caries, bilateral asymmetry and auditory exostoses (see Grauer & Stuart-Macadam, 1998 for further examples; and Hollimon, 2011 for a review of gender studies in bioarchaeology).

Studies of the chemical composition of archaeological bone have demon-strated marked differences in diet between males and females in some past societies (e.g. Privat et al., 2002; White, 2005). Different dietary practices leave chemical traces within bones in terms of the isotopic ratios of carbon and nitrogen (see Chapter 6). This can provide evidence concerning a range of dietary factors including the proportion of protein consumed, whether the diet derived from marine or terrestrial sources and the consumption of certain plant types (e.g. millet and barley leave different carbon isotope signatures).

Such evidence can also provide information pertinent to the life course; for example, breastfeeding and weaning practices and any changes in diet related to age (e.g. Katzenberg et al., 1996; Prowse et al., 2008). Strontium and oxygen isotope ratios within bone can provide information concerning mobility and related cultural practices (e.g. seasonality, exogenous marriage, migration). These chemical components within the bone relate to the geology and climate of childhood residence. Amongst other things, this evidence has been used to examine gendered cultural practices such as matrilocality and patrilocality (e.g. Stojanowski & Schillaci, 2002; Bentley et al., 2005) (see Chapter 6 for more discussion of stable isotopes).

There has been much discussion in the journal *Body and Society* about the gendering of the surface of the body, but gendered practices cut right to the very core of our being, leaving traces in the chemical composition of our bodily tissues and fossilising within the hard tissues of our skeletons. As Budgeon elegantly states: 'Bodies then can be thought not as *objects*, upon which culture writes meanings, but as *events* that are continually in the process of becoming – as multiplicities that are never just found but are made and remade' (2003: 50).

5.2.3   Age-at-death

Following the completion of sex estimation, the next variable of interest for human identification analysis is age-at-death. When dealing with the deceased, this is the age of an individual when he or she died, rather than the length of time since death (the two are commonly confused). Estimation of age-at-death follows two approaches: in younger individuals it involves an examination of the developing skeleton, and in skeletally mature individuals it primarily involves analysing the extent of degeneration in the skeleton (see Figure 5.4 for an example of this from the pelvis). Regardless of the age of the individual, all age estimation techniques attempt to correlate the biological (developmental) age of a skeleton to the chronological age (years) of the person (see Gowland, 2002, 2006; Sofaer, 2006, 2011; Halcrow & Tayles, 2011). No skeletal estimates of age-at-death are perfectly correlated with chronological age; in fact, often they are far from it.

A considerable number of age estimation techniques have been developed for both 'children' and 'adult'[1] skeletal remains, including radiographic, histological and macroscopic methods. An overview of current techniques for estimating age-at-death prior to full skeletal maturity can be found in (amongst

---

[1] Here the terms 'child' and 'adult' are used to differentiate between skeletally immature (i.e. bones and teeth not finished developing) and mature individuals. The authors recognise that these terms are socially constructed and culturally loaded.

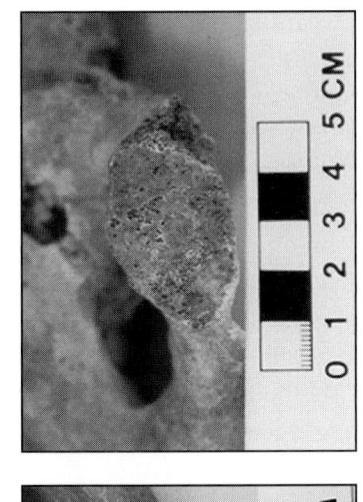

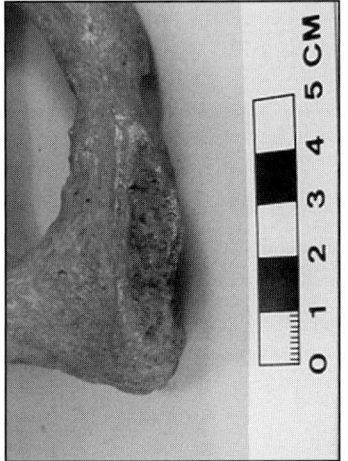

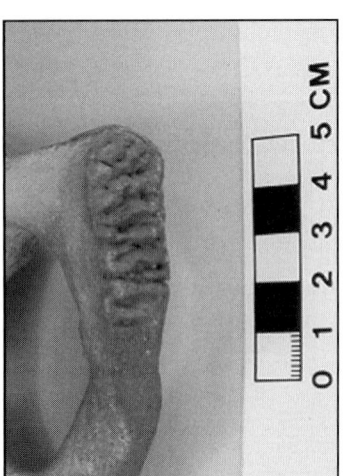

Figure 5.4 Age-related changes to the pubic symphysis from young (left) to old adulthood (right).

others) Saunders (2000), Scheuer and Black (2000) and Lewis (2007). For adult skeletal remains a useful summary is provided by Kemkes-Grottenthaler (2002). Although detailed examination exceeds the confines of this chapter, it should be noted that different populations (e.g. European or African), both sexes and individuals of differing socio-economic status show some degree of variation in the relationship between skeletal maturation and chronological age (see Schmeling et al., 2000; Boechat et al., 2001; Bogin, 2005; Lewis, 2007; Halcrow & Tayles, 2008; Schaefer, 2008).

During growth, individuals can be aged using a range of skeletal and dental techniques, including tooth formation and eruption, long bone growth and the appearance and fusion of ossification centres. Tooth formation and eruption are considered the most accurate and reliable methods of age estimation in 'non-adults' (as skeletally immature individuals are referred to within the bioarchaeological literature) because dental development is strongly correlated with chronological age and only minimally influenced by environmental factors (e.g. disease and malnutrition). Long bone growth (diaphyseal length) is most useful during the foetal and perinatal period (around the time of birth) and into early infancy when growth is very rapid – thus allowing for a reasonable correlation with chronological age. Increasing variability occurs among individuals as they get older. Consequently, after the age of approximately 10 years, long bone growth is a less useful indicator of chronological age in children (Ulijaszek et al., 2000). While the adult skeleton comprises 206 bones, during childhood the number of bones is considerably greater (450 at birth). This is due to the appearance of 'centres of ossification', essentially small bones which develop from cartilage, grow and then fuse on to other bones (Scheuer & Black, 2000). The appearance and fusion of these ossification centres occurs throughout the growth period and can be used to estimate age-at-death prior to skeletal maturity.

The vast majority of practitioners ascertaining age-at-death from skeletal remains rely on macroscopic methods only. This is because they are non-destructive and are no less reliable than the more time-consuming and destructive histological techniques (Aiello & Molleson, 1993). Age estimates of living non-adults have become increasingly important in recent years in relation to the age thresholds for criminal liability, which in many countries is between 16 and 22 years of age (Kellinghaus et al., 2010). This is an issue of particular concern in Europe as European countries increasingly contend with refugees and immigrants who cannot or will not provide accurate or reliable birth dates. Further, as Olze et al. note, the UN Child Convention provides certain rights for those under the age of eighteen, yet establishing this can be problematic (2010). As a reflection of the importance of this work, the Study

Group of Forensic Age Diagnostics was established with the aim of guiding and supporting development and applications in this field. Within this context, skeletal age estimates are often obtained via a radiograph of the hand and dental examination (Schmeling et al., 2006; Schmidt et al., 2008; Olze et al., 2010). Radiographs of the clavicle have also been used (Kellinghaus et al., 2010).

Age estimation for 'adult' skeletons is based primarily on degenerative changes and, as a consequence, substantial inter- and intra-population variation in the rates of ageing exists, confounding attempts to obtain precise ages at death. As with many osteological methodologies, the samples upon which skeletal ageing methods are generated are likely modern in origin, and therefore researchers should be aware of the potential differences in rates of ageing among modern, historic and prehistoric populations (Scheuer & Black, 2000; Usher, 2002). Ageing standards derived from one population may be less accurate when applied to an environmentally distant group, though this has yet to be adequately quantified. Unfortunately, very few population-specific techniques have been developed due to the scarcity of suitable reference material. Therefore, osteological techniques of analysis tend to remain the same irrespective of the origin of the sample being studied.

The most common macroscopic skeletal indicators currently used to derive age estimates from adult skeletal material are as follows: the pubic symphyseal face (Figure 5.4), the auricular surface (both on the pelvis), the sternal rib ends and the sutures of the cranium (see Buikstra & Ubelaker, 1994 or Brickley & McKinley, 2004 for a discussion of the techniques). Numerous studies have examined the accuracy and reliability of these skeletal indicators for determining age-at-death (e.g. Molleson & Cox, 1993; Schmitt et al., 2002; Falys et al., 2006; Samworth & Gowland, 2007; Passalacqua, 2010). Experimental studies have demonstrated that overall the methods of age estimation for adult skeletons are not as accurate as we would wish and comprise largely unknown error margins (Molleson & Cox, 1993; Schmitt et al., 2002). In general, it has been demonstrated that current methods tend to under-age older individuals (Bocquet-Appel & Masset, 1982; Molleson & Cox, 1993). Numerous attempts have been made to counter this bias using different statistical techniques, in particular Bayesian analysis (Konigsberg & Frankenberg, 1992; Gowland & Chamberlain, 2002, 2005a, 2005b; Chamberlain, 2006; see papers in Hoppa & Vaupel, 2002). Such statistical techniques can only aim to portray accurately the degree of variability in human ageing, rather than make the methods any more precise (Gowland, 2007). This is compounded by the fact that the human skeleton does not undergo growth and degeneration in a linear manner, instead one may observe long periods of stasis followed by rapid change. All techniques of age estimation, whether on adult or non-adult skeletal remains, attempt to impose ordinal categories onto what is essentially a variable continuum. Ultimately, human ageing is a highly individualistic process and no amount of methodological tweaking can create a linear, universal phenomenon from it. All that methods can hope to accomplish is a skeletal age that accurately reflects the statistical variability of skeletal indicator stage and associated chronological age within a given skeletal sample.

5.2.4   Skeletal age and identity

Cultural practices can have a profound effect upon the growth and degeneration of the human skeleton (Bock & Sellen, 2002) and are, in part, responsible for the enormous variation in ageing rates that confound attempts to produce accurate chronological age estimates from these remains (Gowland, 2006). Much of the research on age within the human identification literature has focussed on improving methodological techniques. Since the 1990s, however, age as a fundamental aspect of social identity has become the focus of a growing body of research within the social sciences (e.g. Fortes, 1984; Crawford, 1991; Sofaer Derevenski, 1997; Gilchrist, 2000, 2012; Gowland, 2002, 2006, 2007; Sofaer, 2006, 2011; Agarwal & Beauchesne, 2011).

Ethnographic studies have long observed the role of age in the structuring and functioning of past societies (e.g. Mead, 1973; Schildkrout, 1978) Within societies there is often some synchronicity between biological and social age transitions (as with the onset of menstruation) (Myerhoff, 1984: 307); the social identity of a particular age group within a society is, however, *not* the naturalised manifestation of physical development (Beall, 1984). For example, when the ethnographic evidence is examined, bodily maturity is not necessarily a prerequisite for 'adult' status in relation to the skeleton (Beall, 1984). The term 'biological age' is often used to denote the condition of the skeleton or body and is recognised, even in the popular media, as distinct from chronological age. Anthropologists, bioarchaeologists and clinicians will translate these biological, skeletal changes to a chronological age as a more universal unit of analysis. As discussed previously, while it is often conceptualised as purely biological, the manner in which the skeleton ages is the result of an interplay between genetics and an individual's social and physical environment. The skeleton is not a biological entity that exists apart from these influences. As has been observed in studies of sex and gender, the physical and social ageing processes are inextricably linked – each moulding the other. The different definitions of age (see Chapter 2) have been significant in the social sciences because they have enabled the acknowledgement of age as much more than the passing of time or a biological variable. Age is now understood as a key factor in both individual identity and the social structuring of societies (Sofaer

Derevenski, 1997; Gowland, 2001, 2002, 2007; Sofaer, 2006, 2011). Not all academic disciplines conceptualise social or biological age in the same way, and a variety of approaches to the study of age has been adopted. For example, psychology introduces other parameters such as 'behavioural age' (see Sofaer, 2011 for a summary and discussion of different approaches to age within the social sciences).

The field of medical anthropology has produced interesting studies concerning the interrelationship between biology, culture and the life course. For example, amongst traditional Inuit males a hunting lifestyle necessitates a high degree of physical fitness. Once males become adept hunters, however, their fathers reduce their own hunting activities and will then experience disproportionately rapid physical deterioration (Beall, 1984). The physical process of ageing in this instance can, therefore, only be understood within this particular cultural context (Gowland & Redfern, 2010). Old age, as with other age identities, is therefore highly culturally contingent. Even within populations, age interacts with other identities (e.g. gender, ethnicity and status) and these all impact on the way an individual will be perceived (Arber & Ginn, 1991, 1995; Bradley, 1996; Sofaer Derevenski, 1997).

In terms of Western perceptions of ageing and beauty, the facial skeleton has been shown to alter over the life course in such a way that the contours of the face are also affected. Alterations have been demonstrated in, for example, the size of the mandible (Walker, 1995; Shaw et al., 2010; Williams & Slice, 2010). The anti-ageing cosmetic industry has been focussed almost exclusively on the skin, but no amount of botox can prevent such fundamental shifts in the facial architecture. These may precipitate significant changes in appearance which will have repercussions for both social identity and human identification. Interestingly, the cosmetic surgery industry is now modelling these alterations and devising methods of augmenting the bones of older individuals in order to create more youthful-looking facial skeletons (Shaw et al., 2010). In terms of human identification, age-related changes in the skull can cause difficulties when assigning sex to individuals as many of the features affected are those used for sex estimation (e.g. glabella, angle of the mandibular ramus) (Shaw et al., 2010). Walker has discussed this in relation to the skulls of younger adult males which may appear more 'feminine' in appearance than those of older males and which could potentially be incorrectly sexed (1995). Likewise, the skulls of older females may present with features considered generally more masculine.

The physicality of the ageing body plays an important role in shaping social perceptions, but, conversely, the physical body is also culturally conditioned – this is a dialectical relationship (Gilleard & Higgs, 2000; Gowland, 2007). When analysing a skeleton we are observing the impact of cultural factors as well as the passing of time. While chronological time is obviously a significant component in physical deterioration, bodies do not all age in the same way or according to the same timetable. Furthermore, just as different cultures have different perceptions of beauty, age-related physical changes are not universally viewed in a negative light; not all societies place youth on such a pedestal (Gowland, 2007). Ultimately, people grow old within different social and physical environments and these can all have a profound impact on the social and physical experience of ageing.

5.2.5   Ancestry

Determining ancestry using craniometrics (measurements of the skull) has a sinister and contentious history within the discipline of anthropology, associated as it has been with racism, eugenics and Nazism. Since the seventeenth century, cranial indices have been pivotal in the construction of racial groupings and typologies (see Gould, 1997 for a detailed and fascinating discussion). While phenotypic variation in cranial traits among broadly defined population groups may exist, the possibility of harnessing these differences for human identification work is highly problematic due to the specific alchemy of environment, culture and genetics that moulds any one individual's features. In the United Kingdom, many anthropologists abandoned the use of the race concept in teaching physical anthropology due to the arbitrary nature of many of the phenotypic divisions used to differentiate populations and the generally poor correlation between traits and geographical location (Armelagos, 1994). In the United States, however, the determination of 'ancestry' is a common practice within forensic human identification contexts. Indeed, 'ancestry' along with age-at-death and sex is part of the 'Holy Trinity' of anthropological information. Furthermore, in recent years craniometric studies of 'biological affinity' have slowly started to reappear in bioarchaeological research (e.g. Leach et al., 2009).

The term 'ancestry' is now commonly used in place of 'race' in an acknowledgement that humans cannot be assigned to fixed racial categories and in an attempt to distance such studies from the baggage associated with past racial profiling. Genetic and morphological studies show no direct correlation between physical appearance and a person's genetic heritage or place of origin (Edgar & Hunley, 2009; Long et al., 2009; Ousley et al., 2009). Furthermore, as demonstrated by Boas as early as 1912, cranial indices can show considerable plasticity between successive generations due to environmental factors (1912) (see Chapter 2). As discussed earlier, recent research has also demonstrated that the morphology of the craniofacial skeleton changes throughout the life course, and this is likely to have implications for ancestry studies (Williams & Slice, 2010). Consequently, determination of an individual's ancestry is not a clear-cut or simple process and current morphometric techniques are not without their problems (see Keita & Kittles, 1997; Williams et al., 2005).

In spite of all this, craniometric studies have grown more popular in recent years, and a number of methods have now been developed which seek to identify 'population affiliation'. Although craniometric techniques are more sophisticated than they once were, involving complex statistical analyses of measurements using computer software programs (e.g. FORDISC and Cranid), our ability to determine ancestry from the skeleton is still heavily debated (Williams et al., 2005; Ramsthaler et al., 2007). Programs such as FORDISC and Cranid rely on the input of a number of craniometric measurements along with the sex and age-at-death of the individual. These programs measure phenotypic affinity – recognising that certain traits are observed in varying frequencies in human populations in different geographical areas. Proponents of this technique argue that worldwide craniometric variation shows strong geographic patterning (Ousley et al., 2009). Such programs are of course only as useful as the data they are based on. One criticism has been that their reference samples tend to be heavily biased towards particular ethnic groups or comprise a mish-mash of archaeological and modern samples. As a consequence, when used to infer ancestry on skeletal samples not adequately represented within the reference measurements, they have been found unreliable (e.g. Williams et al., 2005; Elliot & Collard, 2009). Similar issues arise in teaching and museum contexts where specimens with 'characteristic traits' tend to be exhibited in order to demonstrate the supposed racial groupings more easily.

Although significant emphasis is placed on the use of the skull for determining ancestry, other parts of the skeleton have also been examined to this end. In particular, the pelvis and the femur have been thought to offer the greatest amount of ancestral variation (Sauer & Wankmiller, 2009). This is not to say, however, that these methods are to be used in the field, as the combined influences of activity patterns, fragmentation and intra-population variation combine to reduce the accuracy of these techniques, often to below a reliable standard (Sauer & Wankmiller, 2009).

5.2.6   Ethnicity and the skeleton

There appears to be confusion in the anthropological literature regarding the relationship between ethnicity and the human skeleton. While the majority of biological anthropologists reject the traditional race concept and recognise that ethnicity is a social rather than a biological construct, they may still engage with the process of determining ancestry (Sauer, 1992).

Consequently there is much debate surrounding the use and validity of craniometrics in anthropological discourse (e.g. Smay & Armelagos, 2000; Ousley et al., 2009). A tension exists between those who seek to identify phenotypic diversity and link it to a particular geographical origin in order to infer population movement, and those who reject any biological involvement in categories of ethnicity. Some of these debates are touched upon by Sauer, who argues for the legitimacy of assigning 'race' to skeletal remains from forensic contexts in the United States, despite recognising ethnicity as a social category (1992). The same author argues for a valid correlation between biology and ethnicity as a social category because of the concordance between social groups and skeletal biology, in particular with respect to the cranial morphology of black and white Americans. This argument does imply, however, a greater ethnic homogeneity in terms of social strata and groupings than is likely the case. Ethnic assignations based on skeletal evidence will likely continue into the future, particularly given the pre-eminence of 'race' within everyday interactions and the perceived biological grounding of this construct. This is certainly still seen in a forensic context, where there is a feeling that skeletal form can suggest race, which in turn can suggest skin colour. Craniometric studies have developed both conceptually and methodologically since the early days of anthropology, but given the plasticity of the human skeleton such descriptive statistics need to be interpreted in terms of the range of human experience and not be reduced to heredity.

Kennedy identifies a two-tier system embedded in discussions of ethnicity and biology, whereby any biomolecular studies of population affinity are tolerated, while the use of morphological traits is seen as racist (1995). However, where contemporary assignations of ancestry differ from past racial categorisations is in the use of the term 'affinity' rather than bounded categories and because there is no sense that these phenotypic differences 'connote fundamental deep differences within the species' (Keita & Kittles, 1997: 534).

Other consequences of the social construction of race are relevant for skeletal analyses, particularly of past populations, and this relates to health. As discussed in previous chapters, race, even though a social construct, becomes biologised through the structured inequalities it creates (Shim, 2005; Gravlee, 2009). For example, studies of post-medieval slave cemeteries in the Caribbean and North America have yielded evidence of high trauma prevalence and skeletal indicators of physiological poor health (e.g. Kelley & Angel, 1987; Rathburn, 1987; Blakey, 2001; Shuler, 2011). All of these are likely the result of poor diet, living conditions and hard labour from childhood through to death – factors that would have had profound consequences on skeletal growth, development and degeneration. Later, we discuss further

the influence of our social environment on health and the subsequent repercussions for the skeleton.

5.2.7   Health, stature and the skeleton

Height is the most ancient form of biometric analysis and was even
included in ancient Egyptian censuses. The height of a living individual is
straightforward enough to measure, but when all that remains is a disarticu-
lated and perhaps poorly preserved skeleton, it is a more complicated affair.
The stature of an individual can be estimated from skeletal remains using two
approaches. The first is the anatomical method, which involves measuring the
height of every skeletal element that contributes to an individual's height in life.
Although extremely time-consuming and dependent on a complete skeleton
being present, this approach is individually tailored and therefore takes into
consideration such things as spinal deformations and differences in body proportions. The method proposed by Fully is the most famous of these approaches
(1956) (though see Raxter et al., 2006 for a more recent revision). It is important
to note that this method provides a skeletal height, not the living height, and
a correction has to be applied to take account of the missing soft tissues. The
alternative method of calculating height is using a mathematical approach.
Here, a given skeletal element is measured and regression equations are used
to extrapolate the living height of a person. These methods are significantly
quicker to use, and allow for stature to be calculated even from highly fragmented remains (e.g. Trotter & Gleser, 1952, 1958). Unfortunately, these methods are based on population averages, and therefore do not take into account
any uniqueness of the body (for example, having a short torso and long legs) or
population variation. Body proportions do vary between populations, and this
is an important consideration when using mathematical methods. Although a
variety of population-specific equations have been produced for people in different parts of the world (e.g. Allbrook, 1961; Radoinova et al., 2002), there are
far from enough studies to cover all modern populations or all archaeological
contexts. In the forensic context, a lack of accurate ante-mortem information
may also make selection of the most appropriate population-specific equations
problematic (Steadman & Haglund, 2005). Furthermore, an age-related correction must also be included since increasing age, especially in females, results
in a decrease in height (Raxter et al., 2006). One last point is the controversy
surrounding the use of the famous Trotter and Gleser stature calculation tables.
Published in 1952, these mathematical formulae allowed for the quick prediction of height while allowing for sex and ancestry differences. They have been
used almost religiously since. Work in the 1990s, which used these formulae
on the same skeletal collection, noted, however, some significant differences

between results that should have been identical (Jantz, 1992; Jantz et al., 1994). Further study concluded that the method described in the original paper was not followed in the creation of those original formulae – of concern was the omission of the medial malleolus from the tibia (Jantz, 1992; Jantz et al., 1994). The implication of this widespread mis-measure of the tibia for such a pro-longed period of time in the forensic and archaeological communities is hard to quantify.

As discussed in previous chapters, in today's Western world there is a strong correlation between social class, health and stature (see Chapter 2). The rate of skeletal growth and final adult height is the result of interplay between numerous environmental factors (e.g. diet, poor health, altitude) and genet-ics. Adverse environmental conditions can prevent an individual from attain-ing his or her genetic potential. Consequently, stature is considered a robust indicator of socio-economic status and is used as an indicator of well-being by, amongst others, anthropologists, archaeologists, social historians and econo-mists (Bogin, 2001). Examples of such studies include the relationship between stature and colonialism, migration, slavery, infectious disease and occupation (see Steckel, 2009 for a summary of stature research). In a 1958 British Birth Cohort Study, height attained at the age of 7 years proved a powerful predictor of unemployment in later life. The chances of unemployment in the shortest fifth of children were three times that of the tallest fifth. In this study, height at 7 years was viewed as a reliable measure of delayed growth as a result of a poor socio-economic and psychosocial environment (Blane, 2006). As Krieger and Davy Smith note: 'both final achieved stature and bodily proportions express the embodiment of nutrition and disease (especially infectious disease) in infancy and childhood' (2004: 96).

Height as a measure of health and the quality of the environment was first used in the nineteenth century by those examining the decline in health asso-ciated with industrialisation and urbanisation (Bogin, 2001). For example, in 1833, Edwin Chadwick instructed commissioners to collect stature data from factory and non-factory children. The findings of his study indicated severe growth stunting of factory children and was influential in precipitating the Factory Act which placed legal constraints on the hours children worked. Chadwick noted that the poor working conditions of children had extremely adverse health and stature consequences. As Bogin writes: 'At 18.5 years of age only the Pygmy populations of central Africa have smaller average heights – about 150 cm for Pygmies versus 158 cm for factory children' (2001: 204). Likewise, Frederick Engels in his *Condition of the Working Class in England in 1844* commented at length on the poor physical state of those working in the industrialised cities in the nineteenth century in relation to poor nutrition and working conditions. In a study of the skeletal remains of children from a nineteenth-century cemetery in Birmingham, Mays et al. found that the cortical thickness of bones was also adversely affected in individuals of lower socio-economic status (2009). Indeed, they suggest that appositional growth may be a more sensitive indicator of environmental stressors than longitudinal bone growth. The relationship between stature and status is a dialectical one; numerous studies have demonstrated that being tall increases the likelihood of upward social mobility; for example tall stature and prospects of promotion at work (see Lindemann, 1999 for a discussion of this phenomenon). As discussed in Chapter 2, our bodily tissues literally come to embody class and status distinctions. In terms of the skeletal evidence, much of this is apparent in delayed growth and stature of poorer classes in the past and present. Additional skeletal evidence is present in the form of so-called physiological indicators of poor health. These include skeletal pathologies such as dental enamel defects (areas of decreased enamel thickness on the teeth) and cribra orbitalia (porosity in the orbits of the skull), conditions frequently observed in past populations and which have multiple causes. A number of studies of archaeological populations has attempted to draw correlations between 'social' status, as ascertained via associated funerary rites, and 'biological' health status as indicated by pathologies such as these (e.g. Robb et al., 2001; Sullivan, 2004; Brickley et al. 2006). Archaeologists are generally cautious to avoid simplistic relationships between social status and skeletal health: elite or high-status cultural practices do not necessarily equate to healthy practices. For example, sun avoidance has been documented for high-status females from a variety of different populations in the past, and this will have had consequences for the vitamin D status of these women (as discussed in Chapter 3, vitamin D is produced in the skin on contact with sunshine) (Brickley and Ives, 2008). Likewise, high-status child-rearing practices in sixteenth-century England, which involved swaddling infants and keeping them indoors, was a contributing factor to the high frequencies of rickets documented for the children of the upper classes during this time (Fildes, 1988). Similar arguments can be made for infants of high-status women in ancient Rome.

As discussed in Chapter 2, socio-economic status and health are strongly correlated in modern-day society and psychosocial stressors play a highly significant role. The latter has rarely been explored within the bioarchaeological literature, and yet many populations in the past had structured inequalities that surely would have created similar stressors. Can we identify psychosocial stress in the bioarchaeological record? The impact of these stresses will be present in the skeleton, but the difficulty, as always, will be teasing apart these from other physiological stressors.

5.2.8   *Traumas and pathologies*

In addition to the determination of sex, age and so on, the osteological profile incorporates other features and events from the life course. These include marks resulting from periods of illness and exposure to physical trauma. These are useful in two ways. From a forensic identification point of view, they are vital for confirming positive identification. This is discussed further later in this chapter, but such identification relies on the comparison of features recorded on the skeleton with those noted and recorded in ante-mortem records. Such records usually derive from the clinical setting. These features of the skeleton can be vital in forensic contexts where other aspects of the osteological profile are essentially the same, such as in war grave contexts like Srebrenica where many of the seven thousand victims were adult males with the same ancestry (Komar, 2003). From an identity point of view, these features are useful because they suggest some event that the person has been exposed to, and this in turn permits comment not only about his or her life but also the individual's social information and practices. For example, certain activities will preferentially expose men or women to particular types of traumas.

Unsurprisingly, a study of traumatic wounds from the skeleton can provide significant information about the context of death, and stab wounds in particular are a very good example of this. For instance, an examination of the kerf (or cut mark) itself can suggest a classification of weapon used in an attack (Bartelink et al., 2001; Humphrey & Hutchison, 2001; Alunni-Perret et al., 2005; Saville et al., 2007; Thompson & Inglis, 2009), while examination of the location of wounds can suggest a positional relationship between attacker and victim (Ormstad et al., 1986; Boylston et al., 2000; Schmidt et al., 2008). This latter point may be vital in some contexts, such as determining execution, genocide and crimes against humanity (as, for example, Warren, 2007 notes in the Balkans), or whether a lethal wound was self-inflicted. The real utility of studying wounds in the skeleton is the fact that their form preserves so well, and often over long periods of time. A fascinating example is the study of the mass grave containing soldiers killed at the Battle of Towton in AD 1461. This War of the Roses battle has been described as the bloodiest on English soil. A study of these skeletal remains provides considerable insights into the weapons used, the type of combat engaged in, the use of amour and information about the soldiers themselves (Fiorato et al., 2000).

Health and social identity are closely linked and, within archaeology, pathological evidence from skeletal remains is considered a vital tool for understanding past living environments, diet and social identities. It is not feasible to provide an overview of all disease types that affect the skeleton (this has been addressed in detail by, amongst others, Aufderheide & Rodríguez-Martín, 1998; Ortner, 2003;

Lewis, 2007; Roberts & Manchester, 2007). On a basic level, bone has a limited response to disease: it forms new bone, destroys bone or a combination of both. Some diseases may induce an osteoblastic response whereby new bone is produced. This new bone is referred to as 'woven bone' and tends to be grey or light brown in colour, and is lighter and more porous than normal, healthy bone. With time, this bone then becomes remodelled into 'lamellar bone', which is denser and mechanically stronger than woven bone and more akin to normal bone in terms of structure and colour. The presence of woven bone on the skeleton indicates that a reactive process was still active at the time of death.

By contrast, an osteoclastic response involves the destruction of bone. A particularly aggressive disease process will result in the rapid destruction of bone with very little reactive new bone formation. For example, secondary metastatic carcinoma in the bone produces multiple destructive lesions with sharp irregular edges. Other pathological conditions result in destructive lesions that have more rounded, sclerotic edges, indicative of bone forming in response to the destruction (e.g. venereal syphilis). The appearance and distribution of skeletal lesions within the skeleton can aid in the construction of a diagnosis. When interpreting skeletal lesions, a 'differential diagnosis' is formulated, whereby several possible causes of the observed lesions are detailed with the most likely cause emphasised. A differential diagnosis acknowledges the fact that many disease processes can result in similar skeletal lesions. Only a small minority of diseases will cause skeletal lesions, and these tend to be long-term, chronic conditions in which the skeleton becomes involved in the later stages of the disease. In addition, diseases will be under-represented in the past as a consequence of poor preservation of skeletal remains.

The impact on health of the transition from a hunting-gathering to an agricultural lifestyle is an interesting example of the kind of palaeopathological analysis conducted on past populations. This transition resulted in a significant dietary shift and increased sedentism, which led to higher levels of non-specific skeletal indicators of poor health, as well as increased dental disease (e.g. Larsen, 1995). Studies of urbanisation and industrialisation have revealed evidence of poor nutrition and health in the form of skeletal evidence for vitamin D and C deficiencies as well as growth stunting in the non-adult population (Mays et al., 2006; 2009). Other such research has focussed on infectious diseases, and the origin, geographical distribution and evolution of these over time (e.g. Roberts & Buikstra, 2003). A large number of studies have focussed on past activities and have examined the utility of skeletal evidence for both joint disease and trauma for indicating these (e.g. Merbs, 1983; Jurmain, 1999). Note as well that primary disease and trauma can lead to secondary conditions presenting (a dislocation may stimulate arthritis, for example). Pathologies and traumas can have a significant effect on mobility and quality of life, which in turn will influence aspects of identity and how they are perceived by others.

5.3   Identifying individuals

Although it is relatively straightforward to create an osteological profile from a skeleton, these are just generalised descriptions of an individual's remains. Actual positive identification of a named individual, usually only required in a forensic setting, is significantly more difficult. In order to do this, one must attempt to match a skeletal feature observed post-mortem to ante-mortem documents which describe or show this feature. The classic example would be the use of ante-mortem clinical radiographs to match fractures and dental records, and this has proven successful. Beyond the use of pathological features for identification in the skeleton, the variations in the morphology of naturally occurring structures have also been used. For example, within the skull can be found a series of sinuses and studies have shown that the specific size, shape and volume of these sinuses are unique to each individual (Smith et al., 2010; Ruder et al., 2011). The pattern of trabeculae bone (the spongy bone within) has also been used in human identification contexts, in a similar way to vascular pattern analysis. Research has demonstrated that not only is the pattern of trabecular bone unique in the bones of the legs and wrist, but that these patterns are visible on radiographs and suitable for unique individualisation (Kahana et al., 1998; Mann, 1998). Potential error could be introduced to this method due to the fact that trabecular bone is subject to remodelling in response to particular stressors and also alters throughout the life course, although work by Kahana et al. on a large sample suggests that stability over time is present (1998). One of the advantages of using bony features such as sinuses or trabecular bone patterns is that they can be examined using traditional radiography, but also by CT imaging which allows for full three-dimensional examination. Such an approach was tested following the 2009 Victoria bush fires in Australia. In February 2009, severe fires swept around the outskirts of Melbourne, destroying whole towns and incinerating over two thousand houses. Because of the nature of the human remains recovered, CT was used to examine skeletal remains inside body bags in order to make assessments of age, sex and unique features for identification purposes, with some success (O'Donnell et al., 2011). Nonetheless, as Grivas and Komar discuss, from an evidentiary point of view, there are still questions to answer regarding the potential error rates of using such unique features in positive identification (2008). This stems from the fact that error rates can be easily created for techniques which study bodily features with finite variation, but less so when dealing with bodily features which have infinite variation (the entire premise of this methodological approach). Christensen and Crowder raise similar concerns, but also add that the discipline must work harder at determining just how unique these 'unique' features are in a given population (2009) (see Chapter 3 for a discussion of similar issues with the use of fingerprint identification).

The other point to consider in this context is the accuracy of the ante-mortem records themselves. First, we must acknowledge that not every country will have high-quality ante-mortem records for its populace. For example, the experience of forensic investigators following the 2004 tsunami in southeast Asia has shown that over 90 per cent of the deceased from Europe had ante-mortem dental records, while the value for Asia was closer to 20 per cent (Petjua et al., 2007). Of course the value of 90 per cent will likely be somewhat inflated compared to Europe generally, since those who can afford to travel to Thailand on vacation are also likely able to afford regular dental appointments. It should also be remembered that simply finding suitable ante-mortem records for this sort of activity can be difficult too. Remaining with the 2004 tsunami, it has been noted that in the initial days following the incident, over twenty-two thousand people were reported missing from Britain alone, a figure which inevitably turned out to be vastly too high (De Valck, 2006). Yet the physical operation involved in the recovery of so many records would have been huge. As it was, by August 2005, two thousand one hundred sixty-seven victims had been identified in Thailand and repatriated to over thirty-six countries (De Valck, 2006). Human identification operations on this scale require many thousands of ante-mortem records from tens of countries to be collated in one location and matched to thousands of post-mortem records. Such an immense logistical task is one reason why identification and ultimate repatriation from such mass fatality incidents takes a long time. Finally, work has shown that ante-mortem records can be incorrect and full of errors. This is true of ante-mortem records from the clinical setting, but also of those provided by family members and loved ones (Komar, 2003). This surely makes comparison with post-mortem records challenging and potentially problematic, and, as already mentioned, makes selecting appropriate population-specific data difficult.

5.4   Conclusion

This chapter has highlighted the dynamic nature of the skeleton and the way in which it may be shaped and moulded by individuals' physical and social environments. The human skeleton has long been characterised as fixed and inert within the social sciences, and yet research has long shown that it provides an important repository for information about individuals and their lifeways. The soft tissues of the body heal and regenerate at a faster rate than bone, thus physical onslaughts as well as chemical signatures relating to diet and mobility may be retained within the harder tissues of the body long after traces have faded elsewhere. This property of bone has been well exploited by bioarchaeologists and biological (and forensic) anthropologists alike. When integrated with other forms of archaeological evidence, the skeletal remains of the people themselves have the potential to provide the most significant and rich contribution towards understanding past identities. In the forensic arena, profiles of age, sex, ancestry, stature and health can be created from the skeleton, which can be crucial for human identification. Within genocide contexts it is frequently the skeletal remains that bear the scars of atrocities on past victims and which can contribute to convictions even many decades after the events. There are limitations inherent to some techniques of skeletal analysis from whatever context, and we have discussed these in detail. Most of these stem from the variable nature of human bodies; we are not easily categorised and scientific methods which seek to do so invariably have to contend with the fact that bodies do not all develop, form and degenerate in a uniform manner. Statistics that attempt to encapsulate or describe human variation are not adequate to the task at hand. Despite these limitations, the skeleton has the key advantage that when the bodily tissues described elsewhere in this book have long decayed, the bones may still be preserved and can help unlock the stories of past lives. Further, the hard tissues of the skeleton and dental enamel provide a unique mineral casing which helps to protect and preserve biomolecular information pertaining to an individual's biography across the centuries. We examine these micro-components of the body in the following chapter.

6

Biomolecular identification and identity

Genes aren't what they used to be.

(Fortun, 2009: 255)

The focus of identity and identification research within the biological and social sciences has traditionally been on macroscopic observations. Technological advances, however, have allowed scientists to investigate the body on a biomolecular scale – the hitherto unknown parts of our physiological make-up – and this in turn has led to a seismic shift in our perceptions of self and others. Naturally, our DNA is likely the first of these microscopic components to spring to mind, but there are others too, including the biochemical composition of our body's tissues and the microbial communities that colonise them. This chapter discusses a few of these biomolecular features of the body in terms of their significance for human identification and identity. The role of DNA for human identification in forensic and anthropological work has already received much attention; less commonly explored within this sphere however, and which we discuss here, is the dynamic interrelationship between the body on this micro scale and social identity and environment. Our DNA is not as immutable or prescriptive in terms of our phenotype as once thought; for example, gene expression can depend on such intangible factors as perceived social isolation, which in turn can have profound physiological consequences (such as poorer health outcomes). Genetic imprinting, epigenetics and social genomics have become a significant field of study and gone are the previously held certainties concerning DNA as a 'blueprint for humanity' (see Atkinson et al., 2009 for a thorough review of the relationship between genetics and society). The concept of the 'triple helix' of gene/organism/environment and the complexity of these interactions in shaping bodily processes and responses has been at the forefront of more recent genetic work (Lewontin, 2000). Despite this, genes are still viewed in a largely simplistic and deterministic way by the general public, and also by many working in the field of human identification.

One of the aims of this chapter is to examine the way in which DNA evidence has been utilised within studies of human identification (relating to past and present contexts) and to indicate possible future directions in the light of current genomic research. In addition, we explore the current significance of DNA, and our understanding of it, in terms of social identity. For example, some have argued that recent scientific advances have resulted in the 'molecularisation' of the body, which in turn has led to the formation of new forms of identity, including the concept of 'genetic citizenship' (Rose, 2001; Heath et al., 2004). We discuss the way in which identities such as gender, ethnicity and class become physiologically embodied at a molecular level. In addition to DNA evidence, the isotopic ratios of particular chemical elements that construct the various tissues of the body have been the focus of a considerable amount of research within anthropology, archaeology and the forensic sciences. While not technically biomolecules, these elements are contained within biomolecules and therefore tend to be subsumed under the title of biomolecular analysis (Brown & Brown, 2011). The isotopic ratios of a range of chemical elements that constitute the body vary in relation to geographical or dietary variables. These too may be influenced by social constructions of identity and we discuss the value of this evidence here. This chapter explores the way in which the environment and society mould the body at a molecular level and the implications of this for both human identity and identification.

6.1   DNA

Deoxyribonucleic acid (DNA) is the biological structure that encodes all development and function within the human body. As a consequence, it is considered of enormous importance in terms of the identity and identification of human beings. Nelkin and Lindee suggest that what most characterises research into the human genome is the acceptance that the gene is the most fundamental component of identity and that by understanding its complexities, we will be able to predict human behaviour and health (1995). However, more recent work within the field of social genomics has shown that the relationship between DNA and a person's phenotype is not a one-way street, and this is discussed in relation to identity in the following section. First, we examine the role of DNA in the context of human identification.

The structure and general function of DNA molecules is described in many publications, such as Sinden (1994), Goodwin et al. (2007), McCabe and McCabe (2008) and Brown and Brown (2011). We present a general summary of these texts here. Simply, a human being or indeed any other eukaryote (a eukaryote being an organism made of cells containing nuclei) has two types of genomic DNA (mitochondrial and nuclear) and so distinct are they that some authors have referred to them as separate genomes (McCabe & McCabe, 2008). The nuclear genome is the form associated with defining human identity, and the typical nucleus in a somatic cell has forty-six chromosomes – half from each biological parent. Twenty-two of these chromosomes are referred to as autosomes, while the remainder are called sex chromosomes. Reproductive and germ cells have half this number of chromosomes. Genetic information is packaged in the famous double-helix form. The helical structure is comprised only of the organic compounds purine and pyrimidine, deoxyribose sugars and phosphate groups. The purine bases are of adenine and guanine, while the pyrimidines are thymine and cytosine. Within the helix, the adenine and the thymine bind together, while the guanine and the cytosine bind. These bindings are referred to as base pairs, and they only occur in these combinations. There are around ten base pairs per turn of this helix, and approximately 3.2 billion base pairs in all twenty-three chromosomes. The chromosomes are not all equal in length, and range in size from around 50 to 250 million base pairs. In reality, the chromosomes themselves can range from fourteen micrometres to seventy-two micrometres in length. Phosphate forms the spine which runs on the outside of the helix. Despite the prominence of the double helix representation within the academic and public literature, many alternative formations and orientations of DNA exist. Such variations include mirrored and inverted repeats. The second, shorter genome is found in mitochondrial cells, is circular and is approximately five-millionths the size of the nuclear genome. Mitochondria are the only organelles in the human body to have their own separate genome (Khrapko & Vijg, 2009). Since, evolutionarily speaking, the mitochondria could be found as separate single-celled bacterial endosymbionts (organisms that live inside another organism), it is not surprising to note that their genome is similar in structure to that of bacterial chromosomes. The sequencing of these two genomes is referred to as 'genomics'.

The double helical structure of DNA was first reported by James Watson and Francis Crick in 1953 while they were working at the Cavendish Laboratory at Cambridge University. Work has continued consistently since then, with the Human Genome Project the largest funded biology project of all time (Allen, 1997). But, as McCabe and McCabe argue, the success of the Human Genome Project does not mean that we understand life or indeed have any sort of blueprint for its construction (2008). Most genes are understood in terms of the biochemical reactions they produce within our cells, and much work is still required to link these to actual phenotypic conditions (Brown, 2001). There are around eighty thousand genes within the genome, separated by extremely long regions of nucleotide sequences, about which relatively little is known (Brown, 2001).

In recent years, a number of social scientists have examined the way in which the genome has moved away from simply a scientific concept to a cultural resource with its own spiritual and moral meaning (Nelkin & Lindee, 1995). Indeed, even the double helix structure of DNA has achieved an iconic status, firmly embedded in the public consciousness (Atkinson et al., 2009). There seems to be a hierarchy at work in terms of the credibility of particular aspects of scientific knowledge (Sofaer, 2006: 35), and DNA is certainly pre-eminent in this respect in terms of public perception; a viewpoint which features strongly in popular television programmes with a 'forensic' subject matter. Indeed Brown and Webster refer to the 'fetishisation' of DNA (2004). The same is true within science itself: projects involving DNA analysis generally attract research funding, feature prominently in 'high impact' journals and are generally considered 'cutting edge' almost irrespective of the application. In relation to the 'race debate' (see Chapter 2), knowledge constructed through DNA carries much greater gravitas and perceived objectivity than the somewhat beleaguered phenotypic characterisations. Within the discipline of archaeology, the study of ancient DNA has long promised the hope of accessing a raft of previously invisible information concerning our ancestors. While results have been confounded by issues of diagenesis and contamination, there is no doubt that important contributions have been made and that current developments (including high throughput 'next generation' techniques) hold great potential.

Within the forensic arena, DNA is perceived to possess great powers of truth divination beyond the capabilities of other scientific realms. McCabe and McCabe note the association of DNA with a 'molecular crystal ball' (2008: 1), and we explore this further in the following section.

6.1.1   *DNA-related aspects of identification*

One of the best known of all of the identification sciences is that frequently referred to as 'DNA fingerprinting'. This technique sprang to prominence in 1988 when used for the first time in the conviction of a murder suspect (Lynch & McNally, 2009). Although often compared to ink fingerprinting, the key difference is the vast amount of information obtainable from a DNA code compared to a fingerprint (Seiden & Morin, 2002). Forensic DNA specialists themselves tend not to use the term 'DNA fingerprinting', as the profiles produced using this technique are not unique (unlike, for example, fingerprints, though see Chapter 3). The analysis of DNA codes for criminal and forensic investigations focusses on a number of loci along the sequence (the actual number varies from country to country); crucially these are in regions of the DNA strand not believed to code for any known phenotype (Seiden & Morin, 2002). Consequently, this region of DNA is often referred to as 'junk DNA'. As Goodwin et al. discuss, if one is trying to use DNA to distinguish one person from another, there is little point in analysing the 99.99 per cent of shared DNA sequence (2007). There are about ten thousand simple tandem repeats (STRs) within the human genome (Brown, 2001) that lurk within the junk DNA and it is these that perform such a vital component of forensic analysis. An STR tends to consist of between one and six base pairs, and commercial STR kits generate amplicons from 100 bp to 450 bp. For example, if twelve STRs are considered in a given forensic case, there is a one in 1,000,000,000,000,000 chance that two unrelated individuals possess the same genetic profile (Brown, 2001). The US and Canadian DNA databases use thirteen STRs, while the United Kingdom databases use ten. As a comparison, there are occasions when German courts will accept just five or six for the confirmation of identity (Bianchi & Liò, 2007).

Single base pair differences have also been used for forensic human identification. Termed single nucleotide polymorphisms (SNPs), they are very prolific within a DNA sequence, yet lack some of the robustness of STRs. These tend only to be used in cases in which the DNA is very degraded, often due to burning, and therefore obtaining a STR profile is problematic. For example, SNPs were used as a last resort for identifications after the terrorist attacks on the World Trade Center of 11 September 2001. One of the key debates at the moment regarding the use of DNA evidence in court focusses on the presentation of the probability statistics. This problem relates to both statistical interpretation and the perception of these statistics. DNA-reporting scientists have to be extremely careful about how they word their findings in this regard, particularly in a court of law. For example, they need to differentiate between the probability that a defendant's DNA matched the crime scene profile and the probability that the defendant was the person who left the DNA at the crime scene (Williams & Johnson, 2008). Misinterpretation of the statistics by prosecutors, for example, can and has led to convictions being quashed.

The principle by which DNA profiles can be used in modern human identification is called the Hardy-Weinberg law. This law states that 'within a randomly mating population the genotype frequencies at any single genetic locus remain constant' (Goodwin et al., 2007: 75). This law makes a number of assumptions: that the population is infinite in size; that mating occurs randomly across the population; that the population is not suffering from the effects of migration; that natural selection is not at work on the population; and that mutations are not occurring (Holsinger, 2001; Goodwin et al., 2007).

Quite clearly, none of these assumptions is absolutely true in the human population. Robust statistical tests, however, have been conducted and significant deviations from the Hardy-Weinberg law have not been detected. We can therefore assume that allele frequencies remain constant in our population and this effectively justifies the use of these statistical methods. DNA evidence as used for human identification in modern criminal contexts should only be used as a means of calculating the *probability* of obtaining the matching profile (the evidence) if the DNA originated from the suspect, from which, on the basis of additional evidence, a verdict of innocence or guilt can be reached (Williams & Johnson, 2008).

One key advantage of the use and adoption of DNA analysis for forensic purposes is that DNA samples can be taken from evidence left at the scene as well as from individual suspects themselves (Williams & Johnson, 2008). Links to personal identity are not reliant on a corporeal presence. In the United Kingdom, samples taken from the scene of a crime can be included on the National DNA Database (NDNAD). The number of subject sample profiles uploaded onto the National DNA Database (NDNAD) increased from 35,668 in 1995–6 to more than 5.8 million in 2011, while the number of crime scene samples uploaded has risen from 2,195 to more than 414,000 during the same period (National Policing Improvement Agency). The efficacy of such databases is currently improving with more advanced methods of sample acquisition and more sophisticated data surveying and comparison, such as with Bayesian networks (Bianchi & Liò, 2007).

With every application of DNA for investigative purposes, whether in the modern or archaeological context, we must remember that analyses are limited by the quality and quantity of DNA recovered. Alaeddini et al. note a number of biological factors that will degrade DNA in the immediate hours after death which result from the failure of cell integrity (2010). These factors are both enzymatic (the result of the release of endogenous enzymes from the body) and non-enzymatic (such as the hydrolytic and oxidative reactions that follow a loss of bodily homeostasis) and can reduce the long DNA strands to just 180 base pairs – often too small a quantity to be of great use in forensic casework. In these instances, SNPs, mitochondrial DNA and a newer method called mini-STRs (e.g. MiniFiler) will be useful. Of course all of this work in forensic investigations is predicated on knowing with an acceptable level of probability that the DNA came from a given person. As such, DNA authentication processes are becoming an increasingly important step in the method (Capelli et al., 2003; Frumkin et al., 2010).

In relation to the NDNAD, it is worth examining what Lynch and McNally refer to as 'biolegality' and the way in which 'developments in biological knowledge and techniques are attuned to requirements and constraints of the criminal justice system, while legal institutions anticipate, enable, and react to those developments' (2009: 284). Biolegality produces 'risky' suspects, 'pre-suspects' and 'statistical suspects'. In the United States, 'John Doe warrants' exist which specify the DNA profile rather than a name of a person 'in order to circumvent statutes of limitation in 'cold cases' (Lynch & McNally, 2009: 284). In the United Kingdom, in order for the NDNAD to become a feasible proposition, legal amendments were necessary to the 1984 Police and Criminal Evidence Act in order to reclassify mouth and hair samples as 'non-intimate', therefore they could be taken without consent, thus allowing the police to obtain samples from anyone in custody. In the United Kingdom, new legislation called the Criminal Justice and Public Order Act has facilitated further expansion of the NDNAD; DNA samples can now be taken from anyone arrested on suspicion of committing a recordable offence. The DNA database also allows familial searching and thus the potential of convicting a relative with a partial DNA match. Familial searching uses the database to track a family member of an offender (related individuals will have more similar STR profiles). The courts should not be able to secure a conviction on this evidence alone and therefore a relative should not be in danger of wrongful conviction based on a partial match. Familial searches are not conducted when very partial profiles are retrieved because there is not enough information and the exercise will generate too many matches.

DNA evidence, with its precise probability and scientific aura, has increasingly been treated as a singular source of an objective truth that should be allowed to overturn verdicts of innocence and guilt based on more fallible 'subjective' forms of testimony and forensic judgement. While important in the forensic context, DNA evidence (both ancient and modern) has also made contributions towards our understanding of our pasts and this will be discussed next.

6.1.2   *DNA and past populations*

Analysis of both ancient and modern DNA of humans and animals has facilitated our understanding of a broad range of past processes and interactions, including human evolution, population mobility, the domestication of plants and animals, and the co-evolution and geographic distribution of ancient pathogens. For a comprehensive coverage of this subject, readers are referred to Brown and Brown (2011). Here we highlight just a few of these developments and discuss their significance in terms of studies of past social identity. The use and potential of ancient DNA in revealing human history has received much attention since the mid 1980s when its use was first reported

(Brown, 2001; Brown & Brown, 2011); this initial excitement and optimism has since been tempered, however, as a consequence of diagenetic limitations and problems of contamination (see Brown, 2001; and Brown & Brown, 2011 for a fuller account). However, over the last 5 years 'next generation' sequencing techniques have produced exciting new results. These techniques allow for the rapid extraction of large quantities of genetic information and have important implications for DNA studies of past anatomical modern humans but also other hominids (Shendure & Ji, 2008).

Studies of DNA have made significant contributions to our understanding of human evolution. Much of this research has relied on modern rather than ancient DNA: the technology (discussed previously) to sufficiently examine the poorly preserved fragments of ancient DNA in terms of evolutionary time scales has not been available until relatively recently. Studies which seek to examine the evolution of anatomically modern humans through modern DNA do so through the examination of modern genetic diversity and a consideration of genetic mutations. Repair and replication of DNA can lead to permanent change in genetic material; change that can range from a single base pair to deletions of large strips of code (Khrapko & Vijg, 2009). To an extent, this change is predictable and is more rapid in mitochondrial DNA. Predictable changes over time allow one to create a form of 'molecular clock' which can be used to study human evolution and migration around the world (Endicott & Ho, 2008). It has been shown that the human DNA sequence has only changed by around 1.2 per cent during the 6 million years or so since we diverged from our common chimpanzee ancestor and that all modern humans share 99.99 per cent of our DNA (Goodwin et al., 2007). The exact causes of these mutations are still the subject of some debate and discussion (Khrapko & Vijg, 2009). Interpretation is complicated by the fact that rates of change vary among species, genes and even sites within DNA sequences (Ho et al., 2005). Additionally, different studies have used varying methodologies for modelling rates of change and data sets (Endicott & Ho, 2008). In the 1980s, the most recent common ancestor for anatomically modern humans (referred to as 'Mitochondrial Eve'), was calculated using a 'molecular clock technique as having lived between one hundred forty thousand and two hundred ninety thousand years ago (Cann et al., 1987). This work supported the 'out of Africa' hypothesis of human evolution, which argued that anatomically modern humans migrated from Africa approximately one hundred thousand years ago, displacing and ultimately replacing indigenous hominid species. The opposing multiregional theory argues that anatomically modern humans evolved from *Homo erectus* concurrently in different parts of the world. In this theory, emphasis is placed on the importance of gene flow between different groups.

As discussed, issues of degradation and contamination have thwarted attempts to recover ancient DNA from archaeological bones. The key influences on DNA survival over long periods of time can be divided into three issues: autolysis, environmental factors and microbial attack (Brown & Brown, 2011). Autolysis is the release of endogenous enzymes following death and the loss of bodily homeostasis. Environmental factors can include radiation and the presence of water, in addition to the effect of heat, which leads to a fragmentation of the DNA (Brown, 2001). The environment, rather than time passed per se, has the greatest influence on DNA preservation, and dry, cold environments are optimum over archaeological time scales (Alaeddini et al., 2010). Microbes are attracted to the biological material after death, and can cause significant damage to biomolecules. Degradation of DNA over time results in the archaeologist being left with very short strands of DNA to analyse, although only 100 to 150 base pairs are required to determine sex, kinship and population affinities (Brown, 2001). Conventional STR typing usually requires 100 bp to 450 bp for success, though mini-STRs are useful when DNA is very degraded and can detect as low as 70 bp in length.

STR methods use Polymerase Chain Reaction (PCR) technology to amplify the short DNA sequence, although the concern with this approach is sample contamination; perspiration, for example, contains more DNA than do ancient samples (Brown, 2011). The damaging impact of contamination can be mitigated through strict controls in both the laboratory and the field (Yang & Watt, 2005; Knapp et al., 2011). The alternative option within an archaeological context is to use mitochondrial DNA. Although shorter to begin with, there are vastly more mitochondria within each cell in a human body.

The successful studies that have utilised ancient DNA have tended to focus on: sex determination, although this is not as accurate as with modern material; kinship studies; analysis of disease pathogens such as malaria or tuberculosis and population studies with particular emphasis on the colonisation of the New World. Note that ancient plant and animal DNA (which can inform about human migration, development of agricultural practices, etc) is also studied using the same methods and experiences some of the same key limitations, although contamination with modern human DNA is obviously less of a problem (Knapp et al., 2011). The recovery and identification of pathogenic DNA from skeletal remains has been undertaken successfully for diseases including tuberculosis, leprosy and bubonic plague. Such studies are useful because only a few diseases result in diagnostic skeletal lesions, therefore studies of ancient pathogenic DNA can potentially lead to increased diagnoses. Willcox notes the ramifications of the use of ancient DNA analysis to prove that tuberculosis arrived in the New World before the arrival of Columbus, thus undermining conventional thought (2002). DNA analysis of tuberculosis is helping scientists to understand the antiquity of this disease and its co-evolution with humans over time (Roberts & Buikstra, 2003). However, studies of pathogenic DNA from archaeological skeletal material can be highly problematic. Success depends in part on the resilience of the structural properties of the pathogen to diagenesis, whether or not the organism becomes sequestered within the bone and contamination from the soil (Bouwman & Brown, 2005; Barnes & Thomas, 2006; Brown & Brown, 2011). As a consequence of these problems of contamination, strict protocols are in place in order to minimise 'false positive' results (Shapiro & Hofreita, 2010).

Another potential avenue for the direct diagnosis of disease from ancient skeletal remains lies in the use of immunological techniques. These seek to identify antigens produced in response to specific pathogens and are frequently used as a diagnostic tool in clinical contexts. Much like blood residue analysis (see Chapter 4), these techniques fell out of vogue in the 1990s due to the high potential for producing false positive results. However, more recently, claims for the successful identification of *P. falciparum* (malaria) from archaeological material using such techniques have been made from soft tissue samples from an Egyptian mummy (Bianucci et al., 2008) and from archaeological bones dating to sixteenth-century Italy (Fornaciari et al., 2010). Debate remains concerning the validity of such results, which are susceptible to diagenetic and contaminant factors (Brandt et al., 2002). There is a resurgent interest in immunological methods (e.g. Mitchell et al., 2008) and with refinements in methodologies (e.g. Schmidt-Schultz & Schulz, 2004) it is possible that such studies will make important contributions to the analysis of past diseases.

Finally, it is worth briefly returning to the potential impact of high throughput 'next generation' sequencing methods. This technique allows the rapid extraction of vast quantities of data even from ancient DNA. Amongst the publicised results has been the sequencing of the genome from the preserved hair of a four thousand-year-old 'palaeo-eskimo' from Greenland (Rassmussen et al., 2010). Proponents of this technique argue that these data may also identify phenotypic traits, potentially allowing us to have fresh insights into the physicality of our ancestors. Naturally this would also be true if applied to modern populations. The interpretation of these data is not without its problems, and issues of contamination and the vast and prohibitive expense of these techniques continually surface (Shapiro & Hofreita, 2010). One of the most exciting applications of this new technology has been the Neanderthal Genome Project, the results of which indicate potential interbreeding between non-African anatomically modern humans and Neanderthals (Pennisi, 2009; Green et al., 2010). Such genetic studies, in conjunction with recent fossil findings, highlight the complexity of human evolution, leading to revisions of previous hypotheses (e.g. Christopher Stringer, 2011, Revisiting 'Out of Africa'). One wonders at the potential implications of such genetic knowledge for our own sense of identity in relation to that of our hominid cousins.

6.1.3   DNA-related aspects of identity

For a number of years, scientists have made strong connections between a person's 'genetic blueprint' and his or her resulting personality, behaviour and identity. In particular, the notion of genetic determinism, that is, that your actions and fate is predetermined by your genes – that people are, as Kirby says, 'nothing more than the sum of their genes' (2004: 185) – has become quite prominent in the popular media. It is worth noting parallels between nineteenth-century perceptions of 'biocriminality', which explored the relationship between criminal propensities, anthropometry and physiognomy and present-day studies, which seek connections between DNA and criminality. If this were true, such direct relationships would have important implications for our sense of self and self-determination. Nelkin and Lindee argue that by emphasising the gene as an absolute deterministic force, one changes the course of public debate about the source and nature of social problems (1995). Emphasis on resolving criminal behaviour, poverty or low education rates therefore shifts away from national or local government, since there is only so much that can be done to combat a person's actions if behaviour is already laid down in the genes. As Allen argues, genetic determinism has a further appeal in that it raises the potential for some technological quick-fix scheme (1997). Kirby also highlights work in this area that suggests a general trend in American society and politics for accepting a 'reductionist view of humanity' (2004: 185). Here, DNA is again used as a political tool while simultaneously forcing the creation and maintenance of categories of identity oriented to our phenotype.

Crawshaw refers to a similar concept as the medicalisation of social problems, that is, removing problems from the social context and placing them firmly into the bodily context (2009). The logical conclusion of this paradigm is that the only way to influence one's fate is to influence one's genes. A development in this area is 'gene therapy'. Gene therapy works by inserting a therapeutic gene into the nuclei of tissue cells using a virus as a vector (McCabe & McCabe, 2008). This allows the tissue cells to synthesise the ribonucleic acid and protein coded by the therapeutic gene sequence. But with so much emphasis placed on DNA being the very definition of who we are, one wonders how altering this affects our perception of self. Wertz asks a similar question when discussing the insertion of neural stem cells into a person's brain (2002). Related to this is the field of pharmacogenomics, which targets drugs and medication based on your genetic code, thereby reducing the possibility of adverse clinical reactions. In addition, there are approximately fifteen hundred genetic tests available that provide predictions about *future* health (Rothstein & Joly, 2009). One must be aware, however, that using genetic codes to determine illnesses or potential problems can also lead to discrimination, for example, in relation to employment and insurance (Brown & Webster, 2004). This is not just a theoretical problem; Lapham et al. questioned 332 members of genetic support groups in the United States to assess the experience of those with genetic disorders of discrimination in the workplace or when purchasing insurance (1996). A quarter of those surveyed believed they were refused life insurance on the basis of their genetic disorder, while 13 per cent believed that they were not given a job, or were fired from their job, as a direct result of their genetic code. Wadman not only highlights the concerns of those with genetic illnesses but also notes that another danger could be that stigmatisation reduces the number of people volunteering for pioneering studies (1998). Part of the problem is that progress is occurring faster in the diagnostic arena compared to the therapeutic arena (Seiden & Morin, 2002). Kirby wonders whether in the future, genetic discrimination will equate to racial discrimination in current Western society (2004).

Perfectly healthy individuals may now be categorised as 'genetically at risk' due to the identification of the presence of particular genetic mutations linked to, for example, breast cancer or Parkinson's disease. The category 'at risk' does not mean that the individual will suffer the disease in question, and yet this identification can seriously affect his or her outlook and identity. Lippman and others have argued that this 'geneticisation' is a form of determinism such that genetic narratives of health and disease tend to shift responsibility for pathology away from the social environment and back to the individual, in much the same way that discourses on genes and behaviour have done (1992). Yet these genetic tests and calculations of risk fail to take into account the significance of gene–environment interactions and epigenetic inheritance which would be almost impossible to quantify with any accuracy (Rothstein & Joly, 2009). In addition, there is a whole spate of other ethical issues surrounding prenatal tests for genetic impairments and the potential abortion of such foetuses (see Scully, 2009 for a review of this issue).

Novas and Rose disagree that geneticisation results in the objectification of individuals, but suggest that it is in fact empowering: 'The rise of the person genetically at risk is one aspect of a wider change in the vision of life itself – a new "molecular optics"' (Novas & Rose, 2000: 485). They argue instead that it has given rise to the new 'somatic self', a development that has important implications for the way in which we manage or 'govern' our bodies on an individual level. Genetic diagnosis enables an individual to adopt a series of life-style changes or interventions that will allow him or her to minimise the risk of developing the disease (Lemke, 2002). Novas and Rose contend that this form of identity is only one of a multiplicity of identities and that it is rarely hege-monic (2000). Individuals construct themselves as carriers of mutated genes (Polzer et al., 2002), and use this information to take charge of their health in a positive way. Genomic medicine redefines the previous medical boundaries between normal and pathological, and instead healthy people can become cat-egorised as 'patients in waiting' (Abu El-Haj, 2007). However, Lock et al., in their studies of genetic risk in individuals with Alzheimer's, found that it played no significant role in terms of embodied identity or family relationships (2007). The lack of certainty in molecular genetics concerning risks of pathology and the plasticity of gene expression means that previous concerns regarding one's genetic identity appears to rest on increasingly shaky foundations (see Clarke et al., 2009 for a review).

While DNA has been held aloft as the essence of who we are and what makes us tick, more recent research has demonstrated that this is not a one-way rela-tionship. We are not predetermined or the sum of our genes; instead genetic instructions are plastic and malleable (McCabe & McCabe, 2008), highly influ-enced by the environmental conditions surrounding individuals and by their unique life histories. Extreme environmental stress acting upon a develop-ing foetus in utero can permanently affect the function and structure of the body (Roseboom et al., 2001), while changes in the environment can alter gene expressions during life (Kuzawa & Sweet, 2009; Roubertoux & Carlier, 2011). Indeed, the term 'imprinting' within the genetic literature specifically refers to the permanent alteration of gene expression due to environmental condi-tions (McCabe & McCabe, 2008). Epigenetic factors refer to those which alter patterns of gene expression while not changing nucleotide sequences of the DNA (see amongst others, Kuzawa & Sweet, 2009 for a description of epigenetic processes). A much cited example of this is associated with the Dutch famine of 1944–5, of which details can be found in Ravelli et al. (1999) and Roseboom et al. (2001). To summarise, in the winter of 1944, Germany halted the ship-ment of food supplies into and within the Netherlands. The result was a severe famine lasting around five months over the course of an extremely cold winter when the average food rations decreased in calorific content from 1,400 per day to 400 to 800 per day. The Dutch famine is significant for researchers as it lasted for a well defined period of time and the identities of those affected are largely known, thus enabling longitudinal studies into the effects of environmental stress. Of particular interest have been the ramifications of this event for chil-dren developing in the womb at that time. For example Ravelli et al.'s study found that females developing in utero at that time had, by the age of 50 years, significantly higher body weight, BMI and waist circumference (1999). Food intake in later life could not explain the differences observed. Therefore, findings suggested that this harsh environment did not 'restrict linear growth but seemed to have resulted in a disturbed central regulation of the accumulation of body fat in later life' (Ravelli et al., 1999: 815). Interesting, it has also been noted that these environmental conditions affected the second generation of offspring as the egg that results in a given person is developing in the mother while she is a foetus in her own mother (McCabe & McCabe, 2008). That political events such as these can have such seismic biological consequences with aftershocks that ripple through the subsequent generations is truly remarkable. There is now a new field of research known as 'nutritional epigenetics' that examines food as a crucial factor in the regulation of gene expression and potentially phenotypic factors (see Landecker, 2011 for a discussion). As in the Dutch famine example, 'This is a model in which food enters the body and, in a sense *never leaves it*, because food transforms the organism's being as much as the organism transforms it' (Landecker, 2011: 177). This is particularly significant for understanding health disparities according to such social categories as race, class and gender, because food consumption often varies substantially according to these variables and could thus result in the perpetuation of embodied inequalities at a molecular level across generations (Kuzawa & Sweet, 2009; Landecker, 2011). As Cole states:

> DNA encodes the potential for cellular behaviour, but that potential
> is only realized if the gene is expressed – that is, if its DNA is
> transcribed into RNA and translated into protein. Proteins shape the
> structure of a cell and determine its characteristic behaviours such as
> movement, metabolism, and biochemical response to external stimuli
> (e.g., neurotransmission). Absent their transcription, DNA genes have
> no effect on health or behavioural phenotypes. (2009: 132)

Genetic studies are starting to reveal some of the mechanisms behind the impact of social environment on mortality and morbidity. Studies have identified specific differences in gene expression in those individuals afflicted with cancer who additionally felt lonely or socially isolated. Interestingly, this affect was related to a *perceived* sense of isolation rather than an objective measure of social contacts. These differences in gene expression were not random but related to three foci responsible for supporting the early phase of immune response (Cole, 2009). Studies have also found that socio-economic status impacts on gene expression (Chen et al., 2008). Social conditions have been shown to regulate the expression of neural genes and significant immune system genes (Cole, 2009). As Cole discusses, 'psychological regulation of gene expression implies that the social world can remodel the functional characteristics of the human body' (2009: 135).

Research in social genomics has now clearly established that our interpersonal world exerts biologically significant effects on the molecular composition of the human body (Cole, 2009). Predictions of genetic risk are therefore highly problematic due to the increasing amount of research that indicates the importance of social environment. Thus 'in connection with virtually all complex diseases … knowledge about genes alone is insufficient to make predictions of much worth about the future' (Lock et al., 2007: 257). So we can say, therefore, that we are clearly more than just an expression of our genes.

6.1.4   Genetic cloning and identity

> Are we witnessing the resurgence of eugenics in new (genomics) clothes? (Abu El-Haj, 2007: 289)

Related to the act of influencing one's own genes is replicating them and creating a new but identical individual. Cloning individuals has long been seen as a significant step in reproductive science and could potentially contribute to the resolution of many issues facing the world today (such as the shortage of organs for transplantation). The actual act of creating a clone is, however, extremely difficult and subject to a very high failure rate. Jaenisch et al. discuss some of these practical difficulties but generally conclude as to the inefficiency of nuclear cloning compared to embryonic stem cell donation (2002). They feel that inadequate programming of the donor material is at fault, resulting in high levels of serious phenotypic abnormalities, so much so that they question whether the surviving clones are normal or merely the least affected (Jaenisch et al., 2002).

Brock disagrees with the general feeling that cloning humans will impact on an individual or society's sense of self and identity (2002). He argues that cloning will produce people with the same genome (and therefore may undermine a sense of genetic individuality), but to assume that this will undermine self-identity is to overlook the significant impact of environment and life history on a person's identity and sense of self-worth. He notes how identical twins have matching DNA but are not 'qualitatively identical individuals' (2002: 314). Neither does cloning undermine the intrinsic importance or value of individual people, since due to the point raised earlier, people do not become simply replaceable. Furthermore, he notes that there is a sense that the significance of life will be lessened if it becomes akin to a manufacturing process, but again he argues that this is not true with IVF children, so should not hold true here. Ultimately he states that most of the fears of cloning regarding its impact on self-identity result from the adoption of a 'crude genetic determinism' (2002: 316), and that the impact on self-identity will result not from the cloning process or from the results of the cloning process but rather from the fears of the effects on self-identity in a form of self-fulfilling mechanism.

If one were to dismiss the creation of human clones, or to borrow from popular culture for a moment, a so-called *Homo sapiens superior*, similar results could be achieved through selective reproduction programmes. The analogy often used in this context is that of breeding with domesticated animals and livestock. McCabe and McCabe directly address the issues surrounding this proposal (2008). Arguments question whether giving a child a genetic advantage in a competitive world is any different from giving a child an environmental advantage, for example, in the form of better schooling. Of course, a key problem is what should the parents do with a child that does not meet the potential that the genetic advantage should provide (an issue also highlighted by Brock, 2002). The fundamental problem to this approach is that most phenotypic traits are likely multigenic and influenced by the environment.

Another term for this approach to reproduction is 'eugenics'. Although today we associate the concept of eugenics with the Nazi regime and the propagation of the Aryan race, it was actually a popular philosophy in the United States and northern Europe in the early twentieth century, in essence growing out of the economic and social situation of the time (Koch, 2009; see Chapter 2). The term 'eugenics' was coined by Francis Galton in 1883, and it referred to the science of improving 'human stock' much like animal husbandry. 'Good' qualities were thought to be inherited, with nature rather than nurture pre-eminent in producing these desirable characteristics (Evans & Schairer, 2009). In the United Kingdom, eugenics legitimated class discrimination whereas in the United States, the distinctions were often along racial or ethnic lines (Evans & Schairer, 2009; Koch, 2009). Here, as Epstein notes, the feeling was that rather than 'permit the Darwinian survival of the fittest to control the gene pool, the object was to ensure the non-survival of those considered to be unfit' (2003: 471). Indeed, as McCabe and McCabe highlight, eugenics became a significant component of state and national policy and law (2008). The US House of Representatives Committee on Immigration and Naturalisation in 1921 utilised a eugenics expert, and consideration was given to the issue during the drawing up of the 1924 Immigration Act. The fear during this time was of an influx of genetically inferior immigrants to the United States following the First World War and their impact on the traditional 'Anglo-Saxon stock'. By 1935, thirty states had introduced sterilisation laws for 'genetically inferior' individuals and over twenty-one thousand sterilisations had been performed; by 1960, that figure had risen to over sixty thousand (Allen, 1997). It is easy to see why this approach had such popular appeal when one notes that many 'undesirable' behavioural traits (such as criminality, poverty, homosexuality and prostitution) were seen at the time as inherited defects. Such associations were born out of earlier work which focussed on detecting the physical features of the 'habitual criminal' (Williams & Johnson, 2008). These associations served to criminalise the body, rather than the criminal action itself, and more specifically with the sterilisation legislation, to further criminalise the female body as 'a site for the production of defective goods' (Nelkin & Lindee, 1995: 24).

Medical geneticists, in their drive to treat debilitating genetic diseases, have been accused of following a eugenic pathway. Both modern medical genetics and older eugenic programmes 'promote exclusion of the disabled rather than inclusion' (Epstein, 2003: 473–4). Interestingly, Epstein at the end of his consideration of the relationship between modern medical screening techniques and the principles of eugenics, concludes that the two are quite different, in part because screening does not seek to improve the qualities of populations. However, by screening for and aborting foetuses with abnormalities we indeed remove them from the population's gene pool. A differentiation is often made between the 'old eugenics' of the past in which state intervention and controls were a primary feature, with the new 'liberal eugenics' of the present in which the state enables the individual to exercise informed choice (Braun, 2007: 11). This notion of informed choice and voluntary consent has certainly been used as a means of legitimising genetic testing and distinguishing it from eugenics of the past. However, distinguishing between old and new eugenics on the basis of consent and an emphasis on the individual rather than the state is perhaps misleading; in the past sterilisation could also be voluntary (Abu El-Haj, 2007; Koch, 2009). Further, even if governments do not have an official eugenic policy, they do often fund screening schemes that fundamentally achieve the same results (Wertz, 2002; McCabe & McCabe, 2008). Indeed many authors find little new in this 'new' eugenics to differentiate it in any significant way to that of the past (e.g. Kerr & Shakespeare, 2002).

6.1.5   *DNA, ethnicity and geographical origins*

Lewontin was the first researcher to apply measures of genetic variability within and among human populations to questions about human 'races' (1972). Work on blood types and biomolecular evidence helped to establish the fact that the range of genetic diversity within population groups far exceeded that observed between groups. Population genetics was pivotal in helping dismantle the Western racial world view, creating a paradigm shift from morphologically static racial groupings to dynamic populations with overlapping gene frequencies (Mukhopadhyay & Moses, 1997: 519). Studies at the DNA level are continuing to provide new insights into genetic diversity within and between human populations (Long et al., 2009). Genetic differences between populations tend to equate to geographical gradients and, therefore, are represented as gradual changes in allele frequency (Goodwin et al., 2007). Human diversity is clinal rather than involving defined bounded groupings, and within any socially defined ethnic group there will be considerable genetic diversity (Klimentidis et al., 2009: 375), and many individuals affiliate with two or more continental groups. Furthermore, the relationship between genotype and phenotype is not straightforward, least of all in terms of the socially complex concept of ethnicity (Klimentidis et al., 2009: 376). While DNA evidence was instrumental in dismantling previous constructions of race as a biological entity, it has nonetheless been used within the forensic context as a means of defining geographical origins; it has become a modern equivalent to measuring head ratios or noting skin colour (McCabe & McCabe, 2008). The Human Genome Project has led to the production of other large-scale databases such as the HapMap project (Bianchi & Liò, 2007), which aims to map genetic similarities and differences across the globe. This has clear implications for population migration studies, but also for forensic investigation where ancestry comprises part of the biological profile created by anthropologists. However, because of the nature of genetic variation, the genetically based ancestry assigned to a person will be both broadly distributed and associated with a large degree of uncertainty (Race, Ethnicity and Genetics Working Group, 2005). As Epstein observes: 'From a genotypic standpoint, population geneticists have concluded that racial categorization has no basis in biology' (2004: 195).

All methods of determining racial groups from DNA information, such as autosomal STRs/SNPs and Y-chromosome STRs/SNPs, make the assumption that the genotypes at specific locations vary randomly across the population. This relates back to the Hardy-Weinberg law. But, as Williams and Johnson note, this is not the case (2008). For example, high levels of within-group mating in some geographic, class and cultural groups in addition to the sharing of alleles amongst families skew the distribution. DNA specialists, however, argue that such subgroups are accounted for through the application of large correction factors that allow for stratification, relatedness and population substructure. Overlooking some of the weaknesses inherent in the Hardy-Weinberg law in the real world can, however, lead to a number of significant social consequences. Kirby, for example, discusses the notion that in the United States, neoconservatives have increasingly adopted genetic determinism and genetic determinations of race in order to enshrine their privileged position in biology, despite the fact that this is flawed from a biological perspective (2004).

Research has also demonstrated the potential use of less obvious genetic markers of geographical origin. For example, work has shown that the frequency of lactase persistence differs from population to population, with 50 to 85 per cent of Europeans and US Caucasians exhibiting the high expression allele while only 5 to 20 per cent of Asians and Africans do (Bersaglieri et al., 2004; McCabe & McCabe, 2008) – although other non-European populations demonstrate high levels of persistence too (Bersaglieri et al., 2004). Lactase is present in young children but decreases in the body after weaning. Without lactase in later life, lactose is not metabolised by the body, and is thus left to intestinal bacteria which produce gas resulting in discomfort and diarrhoea. The cause of this is the differential reliance on dairy products during the past ten thousand years or so, and indeed the high prevalence of lactase persistence matches the distribution of dairy farming (Bersaglieri et al., 2004). Interesting, recent work suggests that the development of lactase persistence in Scandinavia is more recent than this, perhaps less than three thousand years ago (Bersaglieri et al., 2004). Adaptation for effective exploitation of dairy products provides similar selective advantages to resistance to malaria in high-frequency regions of the world (Bersaglieri et al., 2004). This is also clearly another example of the influence of environment on our genetic expression. Genetic diseases are also more or less prevalent from one population to another and this can form an interesting expression of group identity. An example given by McCabe and McCabe is Tay-Sachs disease (a rare and fatal autosomal recessive genetic disorder which affects nerve cells in the brain) within the Ashkenazim Jew community (2008).

6.1.6   Body boundaries and DNA

In our discussion of identity and identification from the skin (Chapter 3), reference was made to the complicated issues surrounding shared body boundaries of conjoined twins. Creating and maintaining boundaries between one another has always been important and, as such, DNA has been proposed as serving as a useful boundary marker. Nelkin and Lindee note that societies commonly draw boundaries to demarcate human identity (1995). This has been challenged recently, however, not just by conjoined twins, but by advances in artificial intelligence and research in animals and primates, that are chipping away at the traditional notions of what made humans unique and special. DNA has thus become a standard-bearer for humanness and its unique status in the world, and is helping to remove 'uncomfortable ambiguities' of identity (Nelkin & Lindee, 1995: 43). Our DNA may serve to differentiate us, but it is not bounded within our bodies; it is disseminated externally through saliva, sweat, hair and discarded skin cells. Within a criminal context, this non-bounded nature of DNA has made it a particularly potent form of evidence. Popular television crime shows have taught us that our DNA surrounds us like a biological aura or dust that is deposited on every surface we touch – both revealing and incriminating. This persistent shedding and externalisation of our inner 'unique code' is a stark challenge to Western perceptions concerning body boundedness.

Although our DNA is so tightly bound with our own sense of identity, it has not been possible for individuals to patent or protect their genome. It does become patentable when a degree of human interaction has resulted in that genome altering in some way. A lack of control or ownership over one's DNA code has led to criticism of biomolecular researchers over the years. McCabe and McCabe note that the Havasupai, a Native American tribe, criticised researchers who had used their DNA, originally donated for diabetes research, to investigate population migration into the Americas; 'the blood of her people had been used to challenge their identity and to refute their religion, and all of this without their permission' (2008: 161). The conflict between a population's genetic history and its oral history and traditions has implications for the sense of identity shared by any given community.

6.2   Bacterial communities

A recent and exciting development within modern forensic science is the realisation of the potential of bacterial and microbial communities to aid investigation. The identification of the species that make up such communities can assist in a number of areas. Initial work has focussed not on the body but on the unique microbial communities within soil and attempts to associate soil recovered from one location with another, in a manner similar to palynology (see Meyers & Foran, 2008 for research in this field, or Gunn, 2009 for a summary). The other key area of research in this field concerns the analysis of microbial agents used in acts of terrorism with the aim of associating the microbes with those individuals who created and nurtured them (Breeze et al., 2005; Jarman et al., 2008). In the current political climate, there is much interest in this type of application, and work such as that by Breeze et al. (2005) draws together many of these themes.

Some work has been conducted on aspects of the human body but interestingly these have focussed on the interpretation of bite marks. To highlight the potential of bacteriology in forensic science, Gunn points out that there are approximately 10^{14} cells in the human body, but that this is significantly outnumbered by the 10^{15} bacterial cells and more than 10^{17} viruses that inhabit it (2009). Bite marks are interesting in their own right as they represent an observable connection between two individuals (Nakanishi et al., 2009). Recovery of DNA from bite marks can be difficult as the enzymes within saliva can denature it, thus leaving only discarded epithelial cells as potential sources (Rahimi et al., 2005). An alternative is to look at the microorganisms deposited in the bite mark. Oral bacteria are transferred from the attacker to the victim during biting, and these can be analysed with techniques such as PCR in order to identify those bacteria present. By far the most common bacteria in such transfers is *Streptococcus salivarius* (Brown et al., 1984), although *S. mutans*, the main bacteria associated with dental caries, has also been detected (Nakanishi et al., 2009). In all, it is thought that *Streptococcus* species comprise of up to 50 per cent of the oral microflora (Donaldson et al., 2010). Crucially for forensic work, *S. salivarius* and *S. mutans* are rarely detected in other body fluids, thus making them very useful for determining bite marks, although they are occasionally found in animals (Nakanishi et al., 2009). Work from the 1980s demonstrated the survivability of oral bacteria; although microbial populations deposited on bite marks reduce logarithmically by around 45 per cent an hour, sufficient quantity was present after six hours to be of use (Brown et al., 1984). Useable microbial DNA was recovered from 99 of 100 blood samples tested, up to ninety-two days after deposition and even in washed samples (Donaldson et al., 2010). Other work demonstrated that these bacteria can survive up to 6 years on stored material, assuming conditions are favourable (Nakanishi et al., 2009).

Experimental work by Rahimi et al. (2005) found that 106 genotypically distinct streptococcal strains were identified from just eight biters. In a similar vein, the presence of streptococcal bacteria, which only reside in the mouth, has been used to distinguish expired blood spatter from impact spatter, since the expired blood will also contain saliva (Donaldson et al., 2010). In terms of the use of microbes in soil profiling, Meyers and Foran state that the key issues determining the effectiveness of the technique focus on microbial uniqueness in different habitat types, the degree of similarity within a habitat and time-related changes to the composition of the community (2008). The same issues can be directed towards the use of microbial communities in and on the human body. There is some interest in the potential of using changes in soil microbial communities to detect the location of bodies. Decomposition heavily involves microorganisms and will likely include many bacterial species, including those from the body itself (Parkinson et al., 2009). This has interesting implications for the calculation of post-mortem interval and time since death, if these changes are also time dependent (Parkinson et al., 2009).

Although they did not fully explore the issue, nor has it been properly expanded upon since, Brown et al. did wonder whether the proportions of various bacterial species in the mouth may act as a form of unique identifier for forensic investigators (1984). Rahimi et al. have made some suggestions following the large numbers of genotypically distinct bacteria recovered from their studies, combined with the fact that some of the predominant strains seem to survive in the biter for several months (2005). Gunn also notes the potential for geographical profiling using microbial communities since some genotypes will vary from country to country (2009). Ikegaya et al. have demonstrated the feasibility of this approach and have applied it to forensic casework (2007). They studied genotypic variation in the JC and BK viruses which both inhabit the kidneys and can sometimes be found in urine. It had been suggested that these viruses could be used to determine geographical origin as they are taken in during childhood and so are a reflection of the region where an individual grew up.

The potential of bacterial communities within forensic and human identification is great but nowhere near full realisation yet. Currently a limiting factor in the development and deployment in the identification sciences is the high technical skill and expensive equipment required to perform and interpret the analyses combined with a lack of full understanding regarding the influences on microbial communities. Nonetheless, these issues will be resolved with time, and we shall no doubt see an expansion of this field.

6.3   Stable isotopes

The chemical composition of an individual's bodily tissues relates to the food and water ingested. A number of key elements that comprise our bodies such as carbon, nitrogen, oxygen and strontium exist as isotopes. Isotopes are different forms of the same element which have the same number of protons but different numbers of neutrons. They are chemically identical but have different atomic masses, resulting in differential fractionation during plant and animal metabolism. Variations in the isotopic composition of the different tissues of the body reflect differences in the type of food eaten (e.g. marine versus terrestrial diet), geology/geography and climate pertaining to both the consumer and the consumed, which in turn are influenced by socio-economic and cultural factors. Therefore, studies of the isotopic ratios of these elements within bones, teeth and even soft tissues have proven of enormous benefit in terms of the study of archaeological populations, but also more recently, forensic cases of human identification. A few illustrative examples of the significance and application of such analyses are provided here.

Isotopic studies of bones and teeth have been widely used in archaeological research, beginning with the seminal paper by Vogel and van der Merwe (1977) and growing in intensity over the last two decades. Initial studies focussed on carbon and nitrogen isotopes. These isotopes are particularly useful for providing dietary information about individuals and groups (e.g. plant types, meat versus vegetarian diet, marine versus terrestrial diet). The carbon isotopes relevant to dietary analysis are ^{12}C and [13]C. The majority of plants incorporate carbon during the fixation of atmospheric CO_2 through one of two photosynthetic pathways (Calvin versus Hatch Slack). The former are referred to as C3 plants (most temperate zone vegetation) and the latter as C4 plants (some tropical vegetation, maize, sorghum, millet). C3 plants have a greater discrimination against ^{13}C than C4 plants and hence have 'lighter' [13]C values than C4 plants. One of the most significant applications of these differences in archaeological research has been in identifying the introduction and/or degree of dietary dependency on maize agriculture in the Americas (e.g. Vogel & van der Merwe, 1977; White & Schwarcz, 1989; Buikstra & Milner, 1991). Maize is a C4 plant and an increased consumption of this foodstuff will result in heavier values of ^{13}C/^{12}C in the bones of the consumers. C3 plants have a more negative $\delta^{13}C$ value (note that isotope values are expressed using delta values: the isotopic composition of the sample is compared with an agreed standard for that element and differences are measured in parts per thousand) with an average value of –26‰ compared to C4 plants with an average value of –12 ‰ (Vogel & van der Merwe, 1977). There is, therefore, a clear distinction in the values produced from C3 plants as opposed to C4 plants. The proportion ratio of ^{13}C and ^{12}C in the diet will also depend upon meat consumption and the exploitation of marine and freshwater resources, which also lead to heavier values of $\delta^{13}C$. In order to aid interpretations, therefore, carbon isotope ratios are usually examined in relation to nitrogen isotope ratios (^{15}N/^{14}N). The proportion of ^{15}N within the body's tissues increases with trophic level, with the bodies of consumers exhibiting ^{15}N values of approximately 3 ‰ higher than their food. Therefore studies of ^{15}N values may indicate individuals with a predominately meat versus vegetarian diet or the consumption of marine resources, which are enriched in ^{15}N compared to a terrestrial diet due to the longer food chains in marine environments (Schoeninger et al., 1983).

Studies of carbon and nitrogen values have revolutionised our understanding of food exploitation and subsistence strategies in prehistoric societies. Food also has significant social and symbolic importance, and studies of stable isotopes have been used to investigate aspects of social identity such as status. For example, amongst the Romano-British cemetery population of Poundbury, Dorset, those buried in high-status graves (e.g. within mausolea) were shown to have a higher proportion of fish in their diet compared to the lower-status individuals (Richards et al., 1998). Likewise Le Huray and Schutkowski's study of skeletal remains from La Tene-period Bohemia demonstrated that males buried with iron weaponry exhibited elevated levels of ^{15}N, indicating a greater meat component to their diet compared to those without (2005).

A recent study of carbon and nitrogen stable isotopes from two Roman towns in Italy dating to the first through the third centuries AD demonstrated high ^{15}N values amongst individuals who also exhibited the pathological condition 'external auditory exostosis', which is often associated with exposure to cold water. The study concluded that, when examined in conjunction, the chemical and osteological evidence indicated that the individuals concerned were likely fishermen (Crow et al., 2010). A number of studies from a variety of sites throughout the world have also indicated dietary differences relating to age, thus highlighting the significance of this aspect of social identity (e.g. Ambrose et al., 2003; White, 2005). For example, at the Anglo-Saxon site of Berinsfield, Oxfordshire, status differences (as inferred by grave goods) were observed in diet, as well as age-related differences between young and old adult males. No sex differences in diet were noted at this site (Privat et al. 2002). At Isola Sacra, a necropolis near Rome dating to the early centuries AD, the collagen of males and older individuals were more enriched in ^{15}N, while the values analysed from younger children and females indicated a more terrestrial-based diet (Prowse et al. 2005).

Another application of $^{15}N/^{14}N$ studies has been the study of ancient infant feeding practices. Infants have been shown to exhibit an increase in $\delta^{15}N$ values compared to their mothers during nursing and then decreasing $\delta^{15}N$ values at weaning. A variety of studies have examined the timing of weaning in surviving infant bones samples. This evidence is important for understanding cultural childcare practices, perceptions of infancy and observed skeletal pathologies and infant mortality (e.g. Katzenberg et al., 1996; Herring et al., 1998; Fuller et al., 2006). Such studies are important for understanding the health and mortality of past populations because infant health has profound consequences for well-being in later life.

More recent developments in stable isotope studies have focussed on those that provide some indication of mobility. Studies of the isotopes of hydrogen, oxygen, lead, sulphur and strontium have been developed in order to reconstruct 'geo-biographies'. The ratios of these elements present in bones and teeth are a reflection of the local geology of the area of residence and climate, particularly in terms of precipitation. Within archaeological studies, the isotopes of strontium ($^{87}Sr/^{86}Sr$) and oxygen ($^{18}O/^{16}O$) have been particularly significant for studies of mobility (e.g. Bentley et al., 2002; Budd et al., 2004; Montgomery et al., 2005). These isotope values are usually analysed from the tooth enamel of individuals. Strontium isotope values relate to the 'bio-accessible strontium' in the area of childhood residence (when the teeth were forming), which relates primarily to the underlying geology, though strontium from other sources (e.g. sea spray) can contribute significantly to the local values (Leach et al., 2009; see also Bentley, 2006 and Montgomery & Evans, 2006 for a review of strontium isotope data and mobility). The oxygen isotope values obtained from tooth enamel relate to precipitation during the time of tooth formation and of particular significance for $\delta^{18}O$ values are climate, latitude and distance from the coast (Chenery et al., 2010).

In theory, it should be possible to identify individual mobility from archaeological skeletons if strontium and oxygen isotope values from the permanent teeth (formed during childhood) are compared to the ratio obtained from the bones (which remodel throughout life) to identify variability and therefore movement between geological regions (Bentley, 2006). Bone diagenesis is, however, problematic when it comes to interpreting both strontium and oxygen stable isotope ratios. Therefore, current studies tend to focus on values obtained from the teeth only. Still, comparisons can be made between teeth that mineralise at different childhood stages to examine mobility and also to identify potential migrant or 'outlier' isotope signals when compared to the range established locally. Indeed, these population level analyses are the most useful for highlighting individuals with anomalous values. This information may then be examined in relation to other osteological features or archaeological evidence such as burial style and artefactual associations to make inferences concerning past mobility (e.g. Budd et al., 2004; Prowse et al., 2007; Leach et al., 2009). Examples of this include a study of Roman York, where researchers identified individuals whose craniometrics and isotope data indicated probable North African origins (Leach et al., 2009). Other studies have employed isotope analysis to examine past cultural practices such as patrilocality and matrilocality. For example, Bentley et al. looked at strontium isotope values amongst a prehistoric population from Thailand and found that men only from this site had a 'non-local' signature, possibly indicating a system of matrilocality whereby men from elsewhere married into farming communities (2005). One important caveat of all of these studies of mobility, of course, is that similar isotopic values can be found in different regions across the globe; so those individuals exhibiting isotopically 'local' signatures could have been migrants from geochemically comparable regions further afield (Bentley et al., 2005).

Within the forensic arena, Stable Isotope Profiles (SIP) have been used for a number of years to establish whether two substances (for example, drugs, explosives, textiles, paints, adhesives) share a common characteristic. While isotope studies on human remains have been conducted within archaeology for a number of decades, only in the last five have they been employed in the United Kingdom to establish identity in a forensic context. One of the first cases related to the torso of an individual that was recovered from the Thames and was referred to by the police as 'Adam'. The circumstances of Adam's death and the material evidence relating to it indicated that the victim had been part of a ritual sacrifice. This provided police with a lead concerning his likely place of origin. Further investigations into Adam's isotopic signature enabled the police to narrow this down to the city of Benin in West Africa (O'Reilly, 2007). Another case is described by Meier-Augenstein and Fraser concerning the dismembered body of a black male from a canal in Dublin (2008). They analysed various bodily tissues including the hair, fingernails and bone, in order to assess the isotopic ratios of carbon, nitrogen, oxygen and hydrogen. Evidence from the hair indicated that the individual had lived in Dublin for at least seven months prior to death as it was consistent with the tap water values from this city. The oxygen values from the bone, however, yielded an isotopic signature indicative of a hot, low altitude coastal region. Further analysis and extrapolation enabled them to estimate that the individual was likely to be a migrant from the Horn of Africa who arrived in Dublin approximately 6 years prior to his death. This evidence provided crucial additional information to the police that facilitated the identification and prosecution of the perpetrator.

This latter example illustrates the way in which different bodily tissues (e.g. hair, nails, bone, teeth) provide chemical information relating to different biographical periods. Nails can provide recent information up to the last six months (likewise hair, though this depends on length: 1cm being equivalent to approximately one month of growth). Mature compact bone can provide information pertaining to 10 to 20 years prior to death, while the cancellous bone sampled from ribs may relate only to the last 3 years. Tooth enamel provides values relating to childhood, with different teeth mineralising at different ages. Furthermore, recent techniques have succeeded in extracting isotope values incrementally from tooth dentine, therefore providing a chronology of chemical values for an individual from a single tooth. An example of this is a study undertaken by Beaumont et al (2012) on the teeth of nineteenth-century Irish migrants buried in London. These represented individuals who were famine survivors as well as two who died during childhood. This record of dietary distress is reflected in the incremental isotope values of the teeth. Such techniques furnish archaeologists and forensic anthropologists with a biographical depth, which in the example of children dying of starvation, charts their deterioration in a dramatic and poignant way.

6.4   Conclusion

Our bodies have been disassembled into their smallest component parts and scientists can now observe the impact of social identity even at this unseen level. Technologically exciting as this is, some believe that identity itself has been reconfigured, becoming geneticised, and humans reduced to mere expressions of their genes (e.g. Lippman, 1992): 'our fetishization of the gene has had the effect of shifting agency away from bodies and towards genes' (Brown & Webster, 2004, 79), while the establishment of DNA libraries has the effect of 'turning the body inside out' (Turney & Balmer, 2000: 411). Steinberg agrees, noting that the 'resurgent biological explanations of identity, social patterns and social relationships, have harnessed new currency in the iconic configurations of the gene and the neo-positivism of recombinant genetic science' (2000: 147). More recent research over the last decade has emphasised the importance of gene/environment interactions and argues against genetic determinism and gene–phenotype causality (Roubertoux & Carlier, 2011). Thus it has been stated that:

> Genes rarely 'determine' phenotypes but instead set the range of outcomes that a biological system may create as it interacts with and responds to the developmental environment. (Kuzawa & Sweet, 2009: 11)

Through our understanding of DNA, we have observed the way in which the bounded body is linked to others, the way in which we not only pass on our biology but also the socio-environmental circumstances of our own conception and development as well as those of our grandparents. DNA evidence has revolutionised human identification within a forensic setting and has led to the redefinition and implementation of laws and governmental policy. Within archaeology, DNA evidence has also had a significant impact, though one currently tempered by diagenetic issues. It is worth stating here that, at present, life at the molecular level is only knowable through the implementation of expensive and elaborate technological processes; it therefore represents restricted knowledge. Isotopic data obtained from skeletal remains are less expensive to recover and have provided a direct means for accessing diet and mobility mediated by social worlds. Bones and teeth fossilise the isotopic by-products of social dynamics in relation to diet and mobility, and thus they have had a considerable impact on our knowledge of past societies. We can conclude our chapter by noting that 'samples of tissue … are far more than simply abstract "bytes" of data. Rather, they are highly corporeal, even visceral, substances loaded with meaning and situated in a nexus of value that can be changeable over time' (Brown & Webster, 2004: 95). Our next chapter reverts back to the macroscopic scale and the external, as we take a final look at the body and the implications of intentional modification for human identity and identification.

7

Intentional modification
of the phenotype

Possibly more so than any other form of identification, body modifications represent the confluence of the biological and social self, and therefore the most obvious intersection of the biological and social sciences. Entwhistle notes that the materials we place at the margins of the body lie on the boundary between self and other, and that these artefacts belong to both our body and to the social world (2002). Body modification incorporates a wide range of activities including: tattooing, piercing, scarification, implantation, tooth extraction, filing and inlaying, body building and anorexia (Featherstone, 1999). It is important to note that most forms of modification involve pain and in so doing the experience or process is firmly located in a corporeal plane (Back, 2004). Intentional modification of the human form has great antiquity, with evidence for tattooing and cranial shaping dating back millennia and from all parts of the world. Within Western culture the body has recently become an increasingly common medium through which personal identity is expressed (or subverted) through a variety of forms of modification; of these, tattoos and piercings are increasingly common. Technological advances over the last century have meant that our ability to modify the human form is immense and growing. More than ever before, the human body is penetrated and merged with inorganic, foreign material, whether for cosmetic, social or medical purposes. The physical form has never seemed so malleable. Indeed, cyberpunk literature talks of 'ditching the meat' altogether – of escaping corporeal limitations through the process of being assimilated and immersed in technology (Poster, 2004).

Much has been written in the sociological literature regarding the motivations and desires behind the practices of modification in the Western world from the prosaic to the extreme (e.g. Featherstone, 1999; Pitts, 2002). For example, Shilling discusses aspects of body modification in relation to the concept of the 'body project' and the shaping of the body as an essential means of constructing the self (1993). Many contemporary accounts of body modification invoke the idea of taking 'control' over one's body (Featherstone, 1999: 2). These debates will not be rehearsed here. By contrast, within the human identification literature, discourse on body modifications has featured very infrequently (though see Black & Thompson, 2007). Of key interest and significance is that body modifications are almost entirely self-directed in terms of the nature, location and timing of their adoption and maintenance (though the modification of children's bodies by adults in the past and present will be discussed later in this chapter). Furthermore, the execution of modifications of this sort is rarely utilitarian (Johnson, 2001). It is this aspect that provides the anthropologist and identification scientist with a potentially powerful tool for understanding and identifying the individual. Body modifications are a largely untapped resource in identification studies and yet, potentially, provide a crucial tool for identifying individuals and groups. Their potential stems from their increasing prevalence, their impact on other methods of human identification (such as visual, pathological or trauma-based identification), the presence of jewellery and associated artefacts within the identification context, the presence of potentially unique identifiers and their application to the identification of the living as well as to the dead (Black & Thompson, 2007). This chapter examines a range of body modification practices and discusses their role in human identity and identification in both the past and the present.

7.1   Dermal modifications

Tattoos and piercings are the most common forms of bodily modification in the Western world. Although often grouped together in studies of modification in the contemporary world, they often have very different aetiologies and motivations. As Table 7.1 shows, such body modifications can be separated into three broad classifications. The significance of this is that it demonstrates the variety in approaches to such modifications. The adoption of these has biases of age, sex, class and so forth and an awareness of this can assist identification scientists in understanding an individual's life. In modern Western populations, dermal tattoos are generally adopted at a younger age by men than by women, although the reverse is true for transdermal piercing (even excluding the massive popularity of standard earlobe piercing).

Brands and tattoos have not always been a matter of personal choice and indeed have been widely used as a denial of individual personhood and as a method of control and surveillance (Connor, 2004; Schildkrout, 2004), for

Table 7.1 *The classification of body modifications*

Classification	Description	Examples
Dermal modifications	Modifications of the surface of the integumentary system	Tattooing; branding; etching; cutting
Sub-dermal modifications	Modifications within the integumentary system, with an entrance wound	Beading; 3d body art
Transdermal modifications	Modifications through the body, with an entrance and exit wound	Piercings; stretching; suspension

example, the branding of slaves and tattoos in concentration camps. Tattoos and body modifications have great antiquity as evidenced by, for example, 'the Ice Man', the preserved body of a man who died in the Alps five thousand years ago (Spindler, 1994). The body of this individual exhibited a series of vertical blue lines on either side of his lower spine as well as on his left calf and a blue cross on his right knee. Some have speculated that these tattoos may have had therapeutic purposes, as degenerative changes were observed in the man's vertebrae. Other direct evidence for early tattooing comes from other arch-aeological contexts in which preservation conditions led to the survival of the skin. For example, both male and female mummies dating to 400 BC from the Altai Mountains (East Central Asia) exhibited elaborate tattoos. The Greeks and Romans seemed not to have practised ornamental tattooing, though there is evidence of slave and criminal branding from 500 BC (Connor, 2004). A religious motivation has also been discussed, amongst, for example, early Christians in Roman territories and later Christian pilgrims to Jerusalem (Back, 2004: 28).

Tattooing in the more recent West is believed to have been introduced after Captain Cook's voyages to Polynesia in the late eighteenth century (Pitts, 2002). Tattoos were briefly popular amongst the fashionable elite before becoming the preserve of more marginal groups. Historically, dermal modifications such as tattoos were considered a sign of belonging to a lower-class subgroup or were associated with rebellion or deviance (Tiggemann & Golder, 2006). Up until as late as the 1960s such groups included sailors, criminals and prostitutes (Gell, 1993; Schildkrout, 2004). Tattoos became a particularly potent way of biologis-ing and stigmatising criminality in the same way that late nineteenth-century scientists associated particular skull shapes and facial features with deviant behaviour (Gell, 1993, discussing the work of Lombroso, 1896 and 1911). This, in turn, led to an element of commonality with so called primitives who, like criminals, were believed to be on the lower rung of the evolutionary scale. By contrast, contemporary taboos surrounding tattoos in modern Western culture, while not completely obliterated, are certainly much more relaxed than they once were and tattoos are now common amongst all social groups and demographics.

Any given body modification is usually a strong reflection of an individual's perception of his or her own identity or the identity that he or she wishes to portray (Black & Thompson, 2007). Extreme forms of body modification are a powerful means of forging new social groups (e.g. gangs), and of expressing resistance and subversive identities. In addition, body modifiers make a statement suggesting that a significant factor concerning this practice involves the notion of 'taking control' of one's own body (Johnson, 2001), since body modifications often involve a considerable investment in terms of time, effort, cost and pain (Tiggemann & Golder, 2006). Pain is integral to many forms of modification and in some cases is an essential component in the transformation of an individual's identity – part of the rite of passage. The contribution of tattoos and piercings to the body and therefore to aspects of identity and identification is complex. This is expressed by Gell in his work on tattooing in Polynesia (1993: 38–9). Tattooing, Gell wrote, is 'simultaneously the exteriorization of the interior which is simultaneously the interiorization of the exterior'. The tattooed skin communicates between the individual and society (Schildkrout, 2004: 321).

It has been argued that the impulse to 'retouch' the body is what distinguishes humans from other primates and animals (Myers, 1992). In terms of personal choice, the stimulus for modifications in most instances in Western society can generally be reduced to one of two propositions: positive self-expression or negative self-mutilation (Braithwaite et al., 2001; Brooks et al., 2003; Quint & Breech, 2005). Beyond this, qualitative surveys (the primary source of data on this topic) have shown that reasons include making an aesthetic statement (Brooks et al., 2003), sensation seeking (including sexual arousal) (Myers, 1992; Millner et al., 2005), as a coping strategy following exposure to psychosocial stressors, to increase attractiveness and sexual appeal, to commemorate a significant event, to increase self-esteem (Myers, 1992; Moser et al., 1993; Johnson, 2001; Roberti et al., 2004; Quint & Breech, 2005; Seiter & Hatch, 2005; Tiggemann & Golder, 2006), to demonstrate individualisation, or conversely to declare allegiance to a specific group (Braithwaite et al., 2001; Brooks et al., 2003). Tiggemann and Golder successfully linked the adoption of biological modifications with the Theory of Uniqueness originally proposed within the social sciences (2006). Their suggestion was simply that modifications were indulged as a consequence of a desire to individualise oneself, within relatively acceptable social boundaries, in an aspect of the modifier's identity and self-concept that they deemed important, that is, their physical appearance. It has also been argued that there is evidence of addictive progression concerning the acquisition of body modifications, perhaps through the influence of simultaneous endorphin release (Roberti et al., 2004) or the process of reconceptualisation following modification that contends that further body modification is now ever more acceptable (Vail, 1999).

Given the generally individualistic and visual nature of tattoos and piercings, they hold clear significance for the identification of both the living and the dead in forensic contexts. In the case of the latter, the dermal–epidermal junction is a significant region for cohesion of the two key layers of the skin (Vioux-Chagnoleau et al., 2006), so when it fails, as it does during the process of decomposition, it allows for the complete separation of the dermal and epidermal layers. This is an advantage when studying tattoos in the context of human identification because with the exposure of the dermal layer, the tattooed image will become clearer until the dermis itself decomposes away.

The relationship between the body and the modification is generally self-evident, given that the modification is placed directly onto or into the body. There are, however, less obvious interactions that can also provide identification information. For example, research has shown that the ink from dermal tattoos leaches from the deposition site into the lymphatic system. From there it journeys into the lymph nodes where the large size of the macromolecules within the ink solution causes them to get trapped. Thus, lymph nodes will be dyed the colour of the tattoo in a process that is permanent, remaining within the nodes even if the tattoo is removed (Dempsey, 2005; Black & Thompson, 2007). It may be possible, therefore, in cases of dismembered corpses, to infer the presence of a tattoo and information about the colour from the appearance of the nodes in the torso. Unfortunately, studies of the lymphatic system have shown that drainage is highly variable among individuals, making it difficult to predict the movement of lymphatic fluids reliably (Reynolds et al., 2007). It is even argued that drainage can cross from one side of the body to the other and that multiple lymph node fields can be used (Reynolds et al., 2007). Some general patterns, however, have emerged from heat mapping studies indicating that, in the arms and torso above the umbilicus, most drainage flows into the ipsilateral axillary lymph node field while below this anatomical feature, drainage focusses on the ipsilateral groin (Reynolds et al., 2007). Observations of lymph node colouration may well in the future offer a sound forensic application in identification techniques. There seem to be no sex or age-related differences in drainage patterns, although there is evidence to suggest that surgery can affect the direction of lymphatic drainage (Reynolds et al., 2007).

These subtle anatomical interactions are not just restricted to dermal modifications. Research has demonstrated that the ring of cells along the edge of a transdermal piercing become necrotic due to the constant movement of the piercing artefact (López Jornet et al., 2004). Therefore, it is argued that it could be possible to age the transdermal modification from the degree of necrotic tissue build-up at the site. Further, piercing items may impact on adjacent tissue structures. For example, tongue or lip piercings may be inferred even in the absence of these artefacts and associated soft tissue structures, by a characteristic wear pattern on either the buccal or lingual side of those teeth that come in contact with it during life. These factors may be significant in terms of human identification in forensic contexts.

7.2   Skeletal modifications

The acts of modifying the skeleton and dentition have attracted the attention and curiosity of archaeologists and anthropologists for many years. Modifications of the skull – cranial shaping – are perhaps the most dramatic of these skeletal alterations. The earliest definite cases of this practice date from late Palaeolithic China and Australia (Pomeroy et al., 2010), though it has been argued that there is evidence from as long as forty-five thousand years ago (Ayer et al., 2010; Prestigiacomo, 2010). The most well-known cranial modifications derive from Peru and Egypt, though examples have been described from many parts of the world. The skull is extremely plastic in infancy and any cranial modifications have to be undertaken during this early stage of the life course. Cranial modification is not the result of a single event, but requires the application of sustained and prolonged force to the cranium and thus a considerable degree of effort on the part of the modifier as well as discomfort, and possibly pain and risk of injury for the subject (Lozada, 2011). As with many skeletal modifications 'it is an irreversible act performed by adults on children' and is therefore a form of ascribed identity (Torres-Rouff & Yablonsky, 2005: 2; Lozada, 2011). The methods of effecting cranial changes are broadly similar throughout the world, and are either tabular (pads used to compress opposing regions of the skull to produce a flattened surface with adjacent bulging) or annular (tight wrappings around the skull to produce a conical effect) (Ayer et al., 2010).

Ordinarily, the growing brain causes the cranial bones to move and osteogenic membranes at the sutures produce new bone to compensate, thus allowing the skull to increase in size and volume. Any consistent pressures applied to the growing skull will affect this normal process; restrictions in particular directions resulting in compensatory growth elsewhere (Aufderheide & Rodríguez-Martín,

1998; Cocilovo et al., 2011). The brain achieves a comparable volume to unmodified crania and there are no apparent adverse effects to the individual with respect to intellectual capacity (Ayer et al., 2010; Pomeroy et al., 2010), although extreme cranial shaping may have an indirect effect on the morphology of the craniofacial bones, altering facial appearance as well as skull shape (Cocilovo et al., 2011). A number of reasons have been proposed to explain such cranial modifications, including intra- and inter-group differentiation, sexual attractiveness and perceived benefits to health (Pomeroy et al., 2010). It has most commonly been employed as a biological signifier of socially constructed ethnicity. As discussed, head shape is a particularly prominent and potent symbol of group identity (Blom, 2005; Torres-Rouff & Yablonsky, 2005). Differences in cranial shape in some societies is thought to have been an indicator of social status (Geller, 2011). For example, Spanish explorers described the Inca elite as modifying their skulls as a means of differentiating themselves from lower-status individuals (Torres-Rouff & Yablonsky, 2005), and skull modification has also been described as a marker of elite status amongst such diverse cultures as Crete's Minoans (Aufderheide & Rodríguez-Martín, 1998) and the Iron Age Sargat culture in the Trans-Urals and Western Siberia (Sharapova & Razhev, 2011). Both males and females were subject to these modifications, though in some cultures particular sexes may be more frequently represented (e.g. amongst the Sargat males' skulls are more frequently modified than females') (Sharapova & Razhev, 2011).

Moulding the body according to cultural norms is a powerful means of constructing intra- and inter-group identities. As Geller states: 'Indelible body modifications speak to a connection between identity constitution and embodied experience. Society's moulding of bodies, in both ritual and quotidian affairs, generates social identity, a forging of historical connections, aesthetic ideals, and future outlooks' (2006). Lozada's study of cranial shaping amongst the pre-Inca Chiribaya of southern Peru noted that the choice of shape was not arbitrary but likely served as a symbolic representation of the surrounding mountains, which were central to their belief systems (2011). Tiesler (2012) reviews the evidence for cranial shaping in Mesoamerica, which was almost ubiquitous in prehispanic populations and highlights the complexity and range of motivations for this practice, not all related to the visible head morphology (e.g. preventing the infant from falling ill).

It is worth noting that in modern societies, similar effects to the tabular process of cranial shaping can occur unintentionally as a result of the sleeping surface and position of babies (Ayer et al., 2010). In the Western world a symmetrical and spherical skull is desirable and products including memory foam and helmets are now available to prevent any deviations from this cultural ideal.

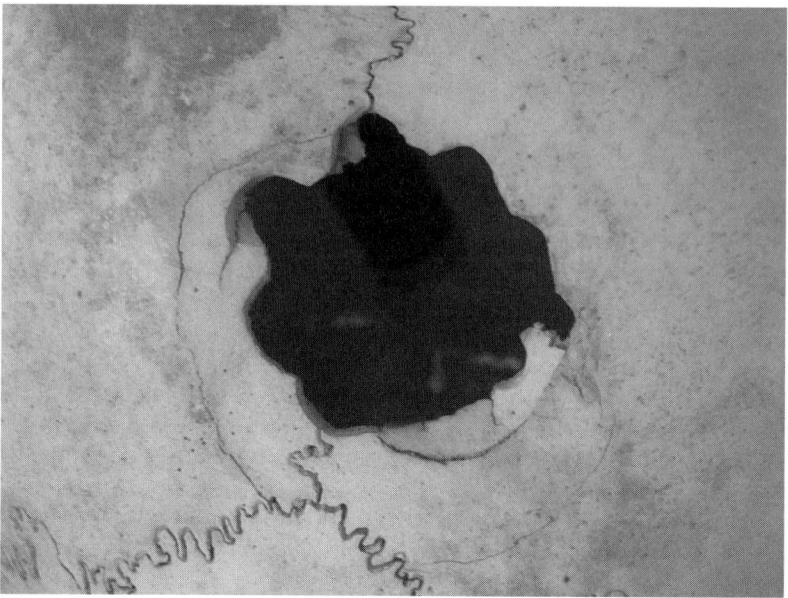

Figure 7.1 The cranium of a non-adult (9 to 12 years) from Peru. The trephination was produced as a result of circular drilling and in this instance appears to be associated with a peri-mortem cranial fracture. There is no evidence of any healing, indicating that the individual did not survive (image reproduced with the kind permission of Dr Catherine Gaither).

While cranial shaping is only successful when applied to the very young, cranial modification in adults can take the form of trepanation – the practice of removing a part of the cranium (Gump, 2010). An example is presented in Figure 7.1. Again, this practice has great antiquity with skulls from as early as the Mesolithic era in Eastern Europe showing signs of trepanation (Aufderheide & Rodríguez-Martín, 1998). It is unlikely that the wound (if the individual survived) would have been highly visible to others and therefore it seems less likely that this practice would have fed into constructions of social identity in the past in the way that cranial shaping may have done. The motivations behind trepanation are likely to have been varied, including ritualistic or medicinal purposes. Some examples of trepanation have been found in association with depression fractures to the skull, suggesting that in some instances the procedure may have been carried out in response to injury or health concerns (Andrushko & Verano, 2008). What is most surprising, given the perilous nature of such an operation, is its apparent success rate in the ancient past. That individuals survived this procedure with apparent frequency is evident by the healed trepanations recovered from the archaeological record – indicated by the rounded, sclerotic bony margins of the lesions in many skulls. Indeed Andrushko and Verano report a survival rate as high as 78 per cent for archaeological examples from the Cuzco region of Peru, indicating well-developed surgical skills (2008).

Dental filing, inlays and ablation are also forms of body modification that have great antiquity and have been observed in many cultures across the world (Ikehara-Quebral & Douglas, 1997; see Milner & Larsen, 1991 for a review). Particularly well recorded are those from Mayan and Inca groups. Some of the oldest known cases have been identified in Mexico dating from 1400 to 1000 BC (Arcini, 2006). Artistic representations of dental modification are depicted on artefacts from this region (Milner & Larsen, 1991). In Mesoamerican populations, both males and females modified their dentition and there is no evidence to suggest that it was status dependent (Milner & Larsen, 1991; Geller, 2006). Dental modification was diminished and eventually quashed during the Spanish colonisation of South America and the introduction of Christianity (Geller, 2006). Dental filing has also been observed in European archaeological populations, although it is not a common finding in this part of the world. Arcini describes filed teeth among some Viking Age male skeletons excavated from what is now Norway and Sweden (2006). Prevedorou et al. also report individuals with dental decoration from an early medieval Islamic necropolis in Spain (2010). In this instance the individuals are thought to have been first generation migrants from North Africa. Amongst modern Western societies the most common form of dental modification is probably the artificial whitening of the teeth, viewed as important for projecting a positive, healthful identity, particularly in North America. Other common practices today include dental inlaying with jewels or the replacement of teeth with precious metals. As discussed in Chapter 5, these hard tissue modifications become vital forms of unique identifiers in many forensic contexts.

The Chinese practice of foot binding also falls within the category of skeletal modification. Originating in around 970 AD and ending in the early twentieth century (Gamble, 1943; Mackie, 1996), this involved the breaking of the toe bones and folding them under the foot to be tightly bound so that it did not exceed three or four inches in length (Mao, 2008). The specific origins of the practice and motivations behind it are unclear; however, it was a means of controlling women by reducing mobility, believed to be aesthetically pleasing and was thought to promote fertility and health (Mackie, 1996). Initially, it is likely to have also been associated with higher-status groups. Adoption was not uniform across China and it seems likely that poorer communities resisted it because it would have limited participation in manual agricultural work (Mao, 2008). As with cranial modifications, this practice was applied to children, usually girls between five and seven; it was extremely painful and complications including ulcers and gangrene could arise, which in some instances could prove fatal (Mackie, 1996; Mao, 2008). As has been noted regarding cranial modifications, because such practices were performed on infants and children these corporeal identities were imposed rather than self-selected.

Since the cessation of the practice of foot binding is fairly recent, evidence can still be seen in elderly woman today. In addition to continuous pain, studies have shown that the effect of foot binding is to increase muscle stress in the lower back, increasing pressure on the pelvis and sacrum (Mao, 2008). The result of these in conjunction with the damage to the feet is to reduce mobility, multiplying the chances of having falls and reduced bone density in the hip and lower spine (Cummings et al., 1997; Mao, 2008).

Although not as extreme as foot binding, the wearing of restrictive footwear in Western societies today has been shown to alter the morphology and biomechanics of the feet (Août et al., 2009). A report from the United States demonstrated that 88 per cent of women were wearing shoes smaller than their feet (Frey et al., 1993). The constrictive effects of high-heeled shoes have long been a concern of medical scientists (Linder & Saltzman, 1998). Mays provides a study of a medieval skeletal population of Britain in which he found a correlation between the lateral deviation of the big toe, indicative of constrictive footwear, and higher social status groups (2005).

In the more recent past, corsetry and its effect on the body, particularly the female form, has received a considerable amount of attention (e.g. Steele, 2001). Evidence of the impact of corsetry on the human body has been preserved in the archaeological record (see Figure 7.2 for an example). Compressed ribs (Figure 7.2) and altered pelvic morphology associated with the practice have been noted from excavated eighteenth- and nineteenth-century skeletons (Klingerman, 2006). Waist binding was practised as far back as the ancient Greeks (Kunzle, 1982), though it was from the sixteenth century onwards that body-moulding clothing was adopted with such zeal in Britain – and even then primarily amongst the upper classes who could afford to limit their mobility (Steele, 2001; Vincent, 2009). In their initial incarnation, corsets were worn by both males and females to create a conical silhouette, but by the eighteenth century they were primarily the preserve of the female, who underwent 'waist training' from childhood onwards. The health repercussions of this clothing were of concern to many medical doctors at the time, though health benefits were also espoused by some (Klingerman, 2006). The corset was pivotal in the gendering of the eighteenth- and nineteenth-century female body; the creation of the 'perfect hourglass' form through the modification of 'normal' morphology was taken to extreme forms and had profound repercussions for female

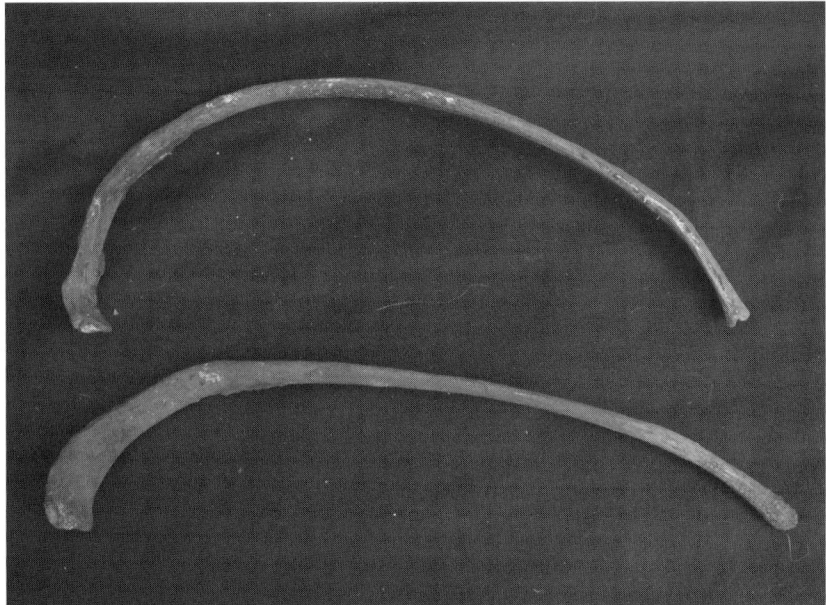

Figure 7.2 A normal rib (above) and one modified due to the application of corsetry (below). Note the flattening of the curve of the body of the rib. Courtesy of Anwen Caffell.

identity, influencing activities, health, behaviour and even fertility. In the nineteenth century, the upper-class female was frequently viewed as delicate in health, hysterical and prone to fainting. It is likely that the severe corsetry adopted during this period was a contributing factor. Thus we observe the dialectical relationship between the corporeal and social construction of femininity. Steele has argued that the female body is similarly constrained today, not by external mechanical pressures exerted on the body by clothing, but rather by internalised forces that seek to create the perfect body through diet, 'body sculpting' exercise and plastic surgery (2001).

7.3   Surgical implants

Surgical implants include a wide range of artefacts such as pins, plates, dental work and soft tissue augmentation. The remit for such implants is also broad, ranging from fracture fixation to aesthetic improvements. Although the range of surgical intervention is wide, orthopaedic surgeons generally leave behind the artificial implants inserted into the body rather than perform secondary operations to remove them (Clarkson & Schaefer, 2007). In terms of human identification, forensic practitioners tend to focus on the potential of surgical implants that have serial numbers imprinted on them. This can lead to identification through comparison with a database. The potential of this in the context of human identification is considerable and increasing in pace with technological development and the number and range of surgical implantations. In 1988, for example, over 11 million Americans had at least one medical implant, while in the United Kingdom, the National Health Service performs around forty thousand hip replacements a year – and the figures will likely rise. Most of these procedures require the implanting of several components, thus providing vast potential for recovering useful implements for identifying both the living and the dead (Ubelaker & Jacobs, 1995; Clarkson & Schaefer, 2007).

Research in this area is limited, with most publications tending to be case studies, such as that of Bennett and Benedix, who used the presence of surgical implants to identify incinerated individuals, a notoriously difficult identification context for obvious reasons (1999). There is a great deal of work still to be done if these artefacts are to be used in a systematic way to identify individuals. While Ubelaker and Jacobs published some manufacturer logos and contact details, they admitted that their database was far from complete (1995). More recently, Clarkson and Schaefer highlighted problems relating to the lack of serial numbers on some implants and the need for a coherent, complete national database (2007). In the absence of information concerning serial numbers, other factors may be of use for identification. For example, not all implants are the same and it may be possible to associate specific implant styles to certain population groups. Different implants are used depending on bone density – a factor strongly correlated with chronological age. For example, Reitman et al. note that traditionally the cementless tapered titanium femoral inserts for hip replacements are preferable for younger individuals, while those with cement and lacking a taper are used for older individuals (2003). Meneghini et al. show that porous materials such as tantalum have also been employed in less dense bone as it has been shown to reduce post-operative reduction in bone density (2010). These are not hard and fast rules, and Reitman et al.'s own work for example, shows that the tapered models are also successful in older people (2003). Furthermore, the style of implant changes over time (for example as demonstrated by McCalden et al., 2009 for knee replacements or Clarkson and Schaefer, 2007 for breast implants), thus potentially giving the investigator an indication of the year of implantation; this may also assist in locating ante-mortem medical records. Likewise changes in implantation methodology may be apparent (for example, Bradley et al., 2009). Different sizes of implants also exist, giving an indication of the overall size of the recipient. The implants may also be constructed from different materials; for example titanium, cobalt/chromium alloy, stainless steel, tantalum, polyethylene and polymers (Clarkson & Schaefer, 2007; McCalden et al., 2009; Meneghini et al., 2010). Different materials may be indicative of patient age (due to their different properties) or the approximate period during which the operation took place (as they may represent an advance in methodology). Likewise different countries and regions of the world adopt different implants and methods, which can also provide useful indications as to the origin of the procedure (Clarkson & Schaefer, 2007). Finally, it should be noted that the presence of a surgical implant cannot be used as a means of positively identifying badly degraded remains as human. Veterinary surgery also employs implants in the care of animals, and some identical devices are used in both humans and fauna (Ubelaker & Jacobs, 1995).

Key to the success of the implant is the interaction of the bone with it. With their review of the success of the cementless tapered femoral implant, Reitman et al. noted endosteal bone formation medial and posterior to the implant, the development of trabeculae bone around the porous aspect of the implant and cortical hypertrophy (2003). This bone formation and remodelling can also be useful in terms of identification as the same authors found these changes to be much more common in the less dense bone of older patients. The issue of increasing age is very significant here. As our population begins to live longer, the likelihood of implant revisions and replacements increases (Meneghini et al., 2010). Detecting whether an implant is a first or second insertion may be difficult, although it may be possible to detect the bone remodelling associated with the first implant.

Breast and muscle augmentation are also increasingly popular. Here generally saline or silicone-gel-filled shells are inserted to alter the topography of the anatomical site. Currently, over 240 styles and eighty-three hundred models of breast implant exist with 1 per cent of the US female population having had them (Clarkson & Schaefer, 2007). Similar to bone remodelling, soft tissue implants can cause associated tissue changes such as capsule development. These can also survive the decomposition process, although may burst following impact or fault.

So far, discussion regarding surgical implants has focussed on its physical interaction with the body and the subsequent implications for identification. As with the discussion in Chapter 4 regarding organ transplantation, there are, however, identity implications of having part of your body replaced with something else, and in this case, something non-organic. A commonality between all of these bodily modifications is that they involve pain. As noted previously, pain is often central to the process of these transformations and perhaps this is seen as integral to the process of identity transformation that accompanies many of these modifications.

7.4   Virtual bodies

Throughout this volume, we have examined the nature of identity and identification from a very physical corporeal perspective. There is, however, an increasing body of work exploring the significance of virtual worlds on aspects of our identity. Such worlds exist online and include popular examples such as *Second Life*. Although it is not our intention to consider this area in great detail, a number of interesting points can be made which are relevant to our overall arguments.

The key viewpoint in this area is that there is an assumption that virtual selfhood differs from actual selfhood (Boellstorff, 2008). This is because to interact in a virtual world such as *Second Life*, users must create an avatar – a representation of themselves within the virtual world. This highlights the main difference between identity in a virtual world and that in the actual world, in that in the virtual world there is a huge emphasis on choice and on the users choosing their appearance. Sutanto et al. (2011) argue that such worlds strip social and physical barriers to communication while Boellstorff (2008) notes that in *Second Life*, the range of options available (which includes choices of age, gender, species) means that avatars are intentionally meant to be interpreted rather than to serve as exact reflections of the self. Therefore, with this freedom of choice comes the opportunity to explore different identities (Stewart et al., 2010), and many virtual world users experience this 'permeable border' between the two worlds in a very positive manner (Boellstorff, 2008: 121).

There is also the view that social identity can be completely severed from the physical identity, something which we argue throughout this book to be impossible. Even in the virtual world, however, there are those, such as Boellstorff, who argue for the embodiment of the physical self (2008). Perhaps the area where this sense of virtual embodiment is most interesting is with disabled users of virtual worlds, where, as Stewart et al. note, 'people with disabilities have a chance to experience life beyond the limitations of their disabilities' (2010: 254). Indeed, the importance of choice re-emerges here as it is the choice of the user as to whether his or her avatar displays any signs of the user's disability. The perception and attitudes towards disabilities differ in virtual worlds too, largely because in the virtual world physical disabilities do not actually exist (they are restricted to the actual world) (Stewart et al., 2010). Indeed research has shown that *Second Life* is beneficial to those with both physical and mental disabilities because it improves their quality of life and their sense of self-worth (Stewart et al., 2010).

Despite this wide range and freedom of choice, much about these virtual worlds are similar to the actual world. The reality for most users is that their avatars do not stray too far from their physical appearance. Indeed, many users describe their avatars as being both themselves and not themselves simultaneously (Hubbell, 2009). This refers to the retention of both physical and psychological aspects of the self on the avatar. Indeed, work has shown that users achieve greater satisfaction from their virtual world usage if they display a more consistent personality and identity to their actual world one (Sutanto et al., 2011). Research has also demonstrated that men and women tend to exhibit their traditional social gender roles in virtual worlds, with women preferring to use *Second Life* to socialise and shop, men to build objects and to own property, younger users for entertainment, and older users for creativity and education (Guadagno et al., 2011; Zhou et al., 2011). Not only can the virtual world influence your expression of identity but your social identity may influence your virtual world usage. Finally, within *Second Life* at least, despite the almost limitless possibilities of the built environment much is a replica of the actual world (Hubbell, 2009).

Interesting aspects of identity ownership exist, such as where one user can have multiple avatars or many users can use the one avatar – thus we have the possibility of multiple identities for one person, or multiple identities in one person. The reality though is a little different, with Boellstorff noting that users tend not to change their avatars' physical appearance much but rather stick to changing clothes and costumes – implying that it is still important to be recognisable even in a virtual world (2008). The fact that a single person has multiple identities is not shocking, and has in fact been referred to and demonstrated throughout this book; however, it might be that in an environment like *Second Life*, this layering of identities is clearer, more defined and more explicit.

There is a difference in the nature of virtual worlds, with some being text based and others utilising 3D graphics, and this has an impact on how aspects of identity and physicality are portrayed and interpreted – largely as a result of how one can express oneself. This means that these different types of virtual worlds hold different functional, experiential and social functions (Zhou et al., 2011). In addition, the 3D worlds allow you to move your viewpoint around, thus giving you a view of yourself in a way not possible in the real world (Hubbell, 2009).

For the users, virtual worlds force them to think about their identities (Hubbell, 2009); perhaps in a way they do not in actual day-to-day existence.

7.5   Conclusion

There are clearly many ways in which an individual can intentionally modify his or her body and appearance. In this chapter we have chosen to examine four broad categories of such modifications, and unlike the preceding chapters, they may encompass a number of layers of the body simultaneously. Nonetheless although they concern different aspects of the body, they do share a number of themes and key is the aspect of self-selection. True, some of the skeletal modifications will have been forced upon children, but tattoos and piercings, surgical implants and other features tend to be voluntary. Thus they are largely about providing other people objects and visuals with which to interpret elements of the modified's identity. This, in turn, results in a very powerful identification tool, whether ante- or post-mortem, since the self-selecting nature of the modification can result in high levels of uniqueness and a close unique and personal relationship with a particular person, which may not be repeated on another. And although the representation of the body in the virtual world is not physical, it seems to be an extension of the modifications that start with the skin and move onwards. Interesting too, is the notion that the topics discussed here themselves fall at the periphery of research in human identity and identification, despite their clear significance in these areas. Nonetheless, as we have seen throughout the previous chapters, aspects of our identities and identification are intrinsically linked to our physical selves. Tattooing for example, produces a 'double skin' or 'artefactual skin' providing a layer similar to clothes, however the tattoo and the skin are indivisible (Gell, 1993: 38). Indeed the bodies created in virtual worlds are literally artefactual, and can be exchanged at will or traded as goods – *Second Life* has a marketplace dedicated to the sale of identities, the implications of which fall beyond this chapter. Changes to the body can be subtle, such as the addition of an earring, but they can also be more substantial, leading to the complete replacement of the physical body by a digital one. Perhaps as significant as the modification themselves is the intention and reasoning behind them. Looking towards the future and the increasing technological advancements within medical sciences and cybernetics, Poster suggests that current body-centred discourses concerning the shaping of the body by society 'appear quaint when confronted by somatic alterations on the horizon' (2004: 87).

8

Conclusions:
identity and identification

The body was never a free gift; it gives temporary shelter to our aspirations on a finite lease.

(Rifkin et al., 2006: 7)

This book has endeavoured to discuss the relationship between the biological and social aspects of our identity and mechanisms of bodily identification. A discursive framework in Chapters 1 and 2 was followed by the dissection of different layers of the body in Chapters 3 to 6, culminating in an examination of body modification in Chapter 7. In each of these chapters, we examined just a few of the multitude of ways in which the biological expression of our corporeal realities is mediated through our social environments. Each layer of the body has been discussed within the context of its specific relationship with body identity and identification, but a number of themes permeate throughout the book and we will draw this volume to a close with an examination of them.

8.1   The construction of identity and identification

The construction of identity is complex, multidimensional, sometimes passive, sometimes active, relational and above all body-mediated whether through individual agency or through the body's capacity to respond dynamically and absorb the by-products of the social fabric. Identity as a concept is difficult to pin down, and teasing out the individual facets of a person's identity (e.g. gender, ethnicity) is harder still, because in essence they are interwoven both biologically and socially. Constructions of masculinity, for example, depend on age and ethnic norms, and each of these facets have biological repercussions on both a macro and micro scale across the bodily tissues. Identification scientists are only just starting to address identity as a social orchestration rather than as a series of easily definable biological categories. There is a clear disconnect between the way in which a forensic practitioner may define identity, compared to, for example, a sociologist. Current body-centred research in sociology has had little impact on the practice of human identification, but conversely, it could be argued that sociologists are just scraping the surface of the multitude of ways in which the biology of the body is saturated with cultural practice.

Throughout this book we have also discussed methods of human identification derived from different biological properties, including fingerprints, DNA and osteoprofiling. We have discussed the ways in which these techniques are culturally ordained, that they are not ahistorical and thus to an extent reproduce our own cultural perceptions about the body (e.g. sex estimation from the skeleton), but that there is also a lack of acknowledgement of this and the possible influence that this may have on interpretations. This is not to say that identification scientists produce 'bad science', but rather that unacknowledged forces exist that serve to create disciplinary realities and products. Scientists do not work in a social vacuum – their work is constructed within the framework of their own understanding, which is context specific. The real interest comes from the interaction of the two: the way in which notions of identity fuel means of identification which then, in turn, crystallise aspects of our identity. The principles of 'DNA fingerprinting' (Chapter 6) serve as a useful example. In this model, our DNA generates our phenotype, which we can then interpret as a variety of aspects of our identity. The realisation that these aspects are important in identification contexts and can be 'read' through DNA fingerprinting has led directly to an increase in interest in this technique. This increase in use has resulted in a feeling that our DNA is the most important determinant of ourselves, and hence has had an impact on how we view identity.

Yet even this is a distinctly simplistic view. Clearly who we are, either physiologically or otherwise, is not pre-written in our DNA; it is forged throughout our lives by the societies within which we live and our position and relationships within this social milieu. As such, 'the body as corporealisation of social and cultural relations has become an increasingly important site of inquiry across a range of (inter) disciplines' (Steinberg, 2000: 148). Throughout all of our discussion of identification methods and in particular biometrics, as we highlighted in Chapter 1, these 'are unique identifiers but they are not secrets' (Schneiner, 1999). This is largely because many of these features are quite visible to the onlooker or investigator. Thus it is something non-biological which grants these biological features meaning in a given identification context.

Finally, the body is a site of more flagrant and purposeful management and manipulations. This has received a great deal of attention within the sociological literature. Giddens discusses the body within late modernity as 'a project', no longer considered a natural given; it is continually open to revision through health and diet regimes, technological interventions and plastic surgery (Giddens, 1991: 218). Bodies therefore become more central to individual identity and objects which can be managed and reconfigured (Budgeon, 2003: 36–7). This is not only a modern phenomenon; within the archaeological record, we also see evidence of extreme body manipulation in the form of cranial shaping, trepanation, foot binding and rib deformation through corsetry (see Chapter 7). While individual agency is emphasised in the literature on contemporary body modifications, for past societies these alterations were associated with group identities (gender, ethnicity and status). Past and present are not so different in this regard; in modern societies group identities certainly exert pressures on individuals to conform to bodily expectations (e.g. gender and status). Further, and specifically within the forensic context, 'the marking of individuals to identify them as criminals – to literally incise a social identity into the flesh – may differentiate such bodies but it does not provide a method to tell one marked body from another. That specific practice, of subjecting a body to examination in order to differentiate it from all others, is founded in a range of techniques that culminate in the nineteenth century "science of identification"' (Williams & Johnson, 2008: 26). Indeed the use of the related evidential paradigm of reading of signs and clues when constructing narratives about the unknowable (also a nineteenth-century construct) still forms the underlying basis for many archaeological and forensic sciences (Crossland, 2009a).

All of this, and the discussions in the preceding chapters, leads us to consider whether there is in fact any objective knowable truth about the body's identity or the manner in which this is identified because the two aspects are so intrinsically linked that they cannot be teased apart.

8.2   The use of the biological and social body

> The individual body should be seen as the most immediate, the proximate terrain where social truths and social contradictions are played out, as well as a locus of personal and social resistance, creativity, and struggle. (Scheper-Hughes & Lock, 1987: 31)

Throughout every chapter in this book we have provided examples of the different ways in which biological and social bodies are co-opted by a range of interested parties in order to address a number of issues. What is of interest is the way in which this very act of using and defining the (biological or social) body in a certain way can actually affect the (social or biological) body. Facets of identity such as race and gender are often seen as social rather than biological constructs, but because they affect the myriad of social interactions an individual experiences they have clear and direct biological consequences. This, in turn, feeds back and influences social identity. Thus an identity feedback loop is created. This loop is particularly apparent with respect to social class, explicitly so in the nineteenth century when working-class bodies were deformed and stunted by gruelling working conditions, poor diet and living conditions, serving to create a biological as well as economic underclass. This biological 'inferiority' was thought to then equate with moral and intellectual inferiority, thus fuelling class eugenics at the turn of the twentieth century. Class inequalities in health and bodily form may be less explicit today, but they are still insidiously apparent; a plethora of poor health outcomes are associated with lower classes along with more obvious phenotypic differences such as higher body mass index. The interaction between the body and class-mediated psychosocial and environmental stimuli help mould the lower class body and mark it apart from the upper classes.

A current example of the co-opting of the biological body for social purposes is the gathering of 'biometric information' for security reasons. Biometric information is now 'surrendered' as a matter of course. Criminals, people applying for passports, welfare recipients, all have to relinquish this information in the name of 'security'. Similar approaches can be seen in the wider medico-legal environment too. In this context, 'the corporeal capturings of biocriminology have shifted from crude body measurements and cuttings to other sorts of specimens' (Walby & Carrier, 2010: 275). It is assumed that this is not an infringement on people, because the body is a natural, static entity, and thus it is perfectly acceptable for this information to be captured, collected and stored indefinitely. We would argue that it is indeed an infringement since through our bodies we are exposed to the world and the recording of our external data, our shape in this world, seems at the same time violating and dehumanising. We are not giving up fragments of our biology, but fragments of ourselves.

Within the medico-legal context we must not forget that the requirements of the identification process are largely driven by the requirements of the legal process itself. The law implicitly assumes uniqueness of an individual, that an individual is bounded and that this can be assessed and determined in some way. The courtroom provides an environment for discussion of the assessment of identity and relies upon the presentation of scientific data regarding various aspects of the physical body in order for this assessment to take place. Our previous discussions on the Daubert rulings (Chapter 3) underline that this is still an important and current dialogue between the legal and scientific communities.

   Also of significance in terms of human identification practices is the intention of the investigator. The human body is subtly but extremely variable.
Identification scientists aim to map this variability, by categorising it in some
contexts, while harnessing this biological specificity in other contexts to create
individualising features. Techniques of analysis therefore either aim to group
individuals, for example according to age and sex categories, or, as is the case
in biometrics, individualise. Identification may be for verification of either the
living or the dead and motives and techniques are context dependent in terms
of the requirement of the 'end user' and the corporeal status of individual (i.e.
dead or living; if dead, the condition of preservation). In fact, much of human
identification is about categorisation, whether we are ticking boxes pertaining
to our ethnic identity for bureaucratic purposes or dealing with archaeological
human remains. The reality of course, is that human biological variation cannot readily be divided into discrete categories to facilitate social interpretation,
however many you choose to employ. Again, attempts to divide individuals
into different ethnic groups serve as a useful example in this regard (for more
detailed discussion see Chapters 2, 3 and 4). As Relethford observes: 'in purely
statistical terms, the cultural construction of race is the transformation of a
continuous variable into an ordinal-level or nominal-level variable with the
attendant loss of statistical information' (2009: 21). Furthermore, phenotype is
not equivalent to genotype precisely because observed traits are a function of
gene expression and not simply gene frequency: 'To assume that phenotypic
variation among humans is a function solely of inherited genes is an ideological
argument not a scientific one' (Krieger, 2005: 2157). Despite evidence suggesting a lack of biological basis for racial groups, the fact that there is a perception
that they exist does in itself influence biological factors in a similar feedback
loop to that mentioned in relation to class. For example, differences between
ethnic groups in terms of both morbidity and mortality have repeatedly been
documented in the United States, the United Kingdom and elsewhere (Nazroo &
Williams, 2006: 238). The factors underlying such differences, however, remain
contested. An adequate understanding of racism is fundamental to understanding the social inequalities in health. As Gravlee discusses, to critique race is not
to deny the range of human biodiversity or the relevance of genetics (2009: 51).
Instead the concept itself is simply not adequate to describe the complexity of
human variation. One can make similar arguments in relation to gender, age
and other corporeally tethered categorisations, including disability. And yet,
we require such categorisations; they are not simply a social indulgence, they
fulfil disciplinary and societal functions. They are important discursive devices
that help shape our social realities, but also describe them in ways meaningful to us. This is not to say that these categories have no empirical basis or are

simply corporeal imaginings. A bioarchaeologist can (usually) differentiate the skeleton of, for example, an older adult female from a young adult male. What is important to remember and what identification scientists do not always acknowledge is that the interpretation and boundaries of our categorisations are not fixed, but are subject to shift. Again this is most obvious in relation to ethnicity, but also to age and gender. Further, the social meaning of being a male or female within any one society leaves different biological traces interwoven with other culturally meaningful forms of identity. Remnants of these experiences are recorded in different biological tissues pertaining to life course stages from childhood (teeth) through to the present (hair).

The scalar nature of the body is another theme that has emerged within this book. When we consider the body, we consider the individual, but also in relation to society, and then broaden this out to populations on a continental and global sense. Likewise, when we consider the biological body, we do so at a micro scale with respect to DNA and stable isotopes, we consider individual organs or tissues from the skin to the skeleton or we take into account the entire human form. The body similarly exists both biological and socially on different temporal scales. There is a tendency within archaeology to view the body (skeleton) as representing a snapshot of an individual in death. As discussed earlier in this chapter and in Chapters 2 and 6, the tissues of the body (including the various hard tissues) all have different life spans and will therefore encompass different biographical components of an individual's society–body interactions from childhood to old age (Robb, 2002). However, beyond this, an individual's gene expression and thus their response to stimuli is altered by the social environment of his or her grandparents. Therefore, via our DNA, the socio-economic circumstances of recent ancestors may reverberate across generations so that our tissues are not only a product of our own biographies but interwoven with those of our forebears, affecting our bodies, immune response, health outcomes and phenotype. Thus it has been argued that health inequalities amongst black Americans stems in part from the social imprinting of the genes of their slave ancestors (Jasienska, 2009). Obviously, we can extend this still further, to argue for this interrelationship on evolutionary time scales – our bodies and responses are the product of these social and physiological interactions over the one hundred thousand years or so of anatomically modern human existence and beyond.

The temporal aspect of the body also comes into play with regards to the identification of deceased individuals. Preservation and decomposition of the body's tissues are partly time dependent. After death, the tissues of the body decay at different rates and it is the hard tissues and biomolecular remnants within that generally remains across the centuries, providing echoes of a life lived for archaeologists to then interpret. Individuals continue to have social agency beyond death in the memory of their loved ones, but their physicality may persist too and the body itself can become an important social agent in some circumstances even millennia after death. For example, the recent calls for the reburial of Neolithic burials at Avebury in the United Kingdom by modern-day druids have led to a lengthy and expensive consultation process and generated a great deal of media and public interest. But the decayed body in itself must be made understandable in some culturally specific way. So with the mummified body of the man recovered from the Alps (referred to as the Ice Man or Ötzi), Robb argues that:

> giving the Ice Man a name, thus, is anything but coincidental; it is an essential part of repersonalizing this body. The miraculously preserved 'glacier mummy' is a purely material trace. But bodies are social, and when a body is presented, anomalously, outside of social relationships, those finding it have to extemperize, to use constructional reflexes provided by their society to make good the missing sociality. (2009: 122–3)

In the current climate of the use of biometrics in relation to national security and the scientific dominance of the 'gene', another theme that has emerged has been one of biological determinism. Not so much one body influencing the other, but of the biological body controlling the behaviour and actions of the social body. This is particularly noticeable within genetics studies. In the clinical context, Steinberg notes that 'genetic science produces accurate facts; doctors explain these facts to patients, who are in turn grateful; and 'rational steps' can then be (and are) taken to accommodate the genetic propensities revealed' (2000: 163). However, the use of biological determinism does not stop in the GP's office and the difference between eugenics of old and the 'new eugenics' has not adequately been differentiated. Likewise, a recent resurgence of craniometric studies to identify ethnic groups or 'biological affinity' within forensic anthropology and human bioarchaeology suggests a return to the old ways; repackaged, incorporating more sophisticated technology, but still not adequately theorised or differentiated from past practices. Beyond this there is great interest in the use of DNA profiles to predict or explain criminal behaviour with a 'resurgence in discussion of biocriminology in textbooks after a period of taboo from 1930s' (Wright & Miller, 1998). We may have shifted from descriptions of low foreheads and close-set eyes as predictors of criminality to genetics, but these are all physiological descriptors that despite being interpreted in isolation from social context are used to interpret deviant social actions.

As Brown and Webster discuss: 'the informaticized (and geneticized) body has become subject to (digital and genetic) codes that, though reconfiguring the body and its lived experience, do not thereby always work towards its "emancipation"' (2004: 51).

While recent research has emphasised the importance of social context in terms of 'gene expression', the spectre of biological determinism in relation to genetics does seem to have a profound political currency and popular salience. Bodies are being reduced to and made 'knowable' or at least 'identifiable' through numerical codes and one wonders about the repercussions of such dehumanising technology. The Human Genome Project was also an (expensive) attempt to map the human body and make it knowable. This project has not brought us any closer to understanding 'the essence of humanity', to knowing who we are, nor has it resulted in any significant developments in our understanding of how to cure diseases. This is because we are more than the sum of our genes; we are the products of our social and physical environment. It is this very plasticity that is the source of our variability and adaptability.

8.3   Issues with the disciplines

It is clear from the examination of the literature in the preceding chapters that part of the difficulty in fully exploring this subject stems from the disciplinary divisions that fall on either side of this nature–culture divide. There is a lack of integration of the biological and social literature concerning the body, in addition to a lack of full and proper integration *within* individual disciplines (e.g. archaeology, anthropology). This is further compounded by the manner in which the physiological body has traditionally been approached. As mentioned in Chapters 1 and 2, previous volumes on this topic have tended to focus on one particular biological element, such as the skeleton or the skin. The artificial and false categorisation of the body, while making comprehension easier, fails to realise fully the complex and interrelated components and systems of the body. This is something that we have sought to avoid as much as possible when writing this book, although we acknowledge that we have divided our chapters in such a manner.

Ultimately, biological and social bodies are enmeshed in ways we cannot begin to unravel. The more one examines the body in society, the more one notes the false and unsustainable dichotomy of the nature–culture divide. These artificial divisions restrict collaboration between workers and integration of the literature and theories, are thus self-sustaining and debilitating. As Scheper-Hughes and Lock discuss, the separation of the mind and the body has been rarefied in the vocabulary of scientific discourse: 'We lack a precise vocabulary with which to deal with mind-body-society interactions and so are left suspended in hyphens ... forced to resort to fragmented concepts as the bio-social, psycho-somatic ... as altogether feeble ways of expressing the myriad ways in which the mind speaks through the body, and the ways in which society is inscribed on the expectant canvas of human flesh' (1987: 10). We hope that this book is a small contribution towards integrating some of this literature in relation to the practice of human identification.

8.4   Human identity and identification

The human body is a complex entity, both from a biological and social stance. Despite centuries of study, we still lack a full understanding and continually debate how we move from conception to death, both physiologically and socially. We have made an attempt to address these issues in this volume, partly by not shying away from this complexity. Thus it was our intention to draw together our discussions of the constructs of human identity and identification throughout the layers of the body. And while a number of interesting points have been highlighted, more general themes run across these discussions: the difficulties of separating the influence of both the biological and social world on identity and means of identification, the way in which these constructs are used by other people and how that feeds back on the individual and the issues of studying the body from the perspective of so many seemingly disparate academic disciplines.

We have consistently argued throughout this book for the significance of the biological body in determining social aspects of our identity and identification and vice versa. The processes of human identity and human identification will always demand that many dynamic elements be converted into static data. But the multidimensional qualities of the body as discussed throughout this book must be readily acknowledged in order for techniques of identification to be furthered. Finally, as Nelkin and Kindee note, 'biological differences in themselves have no intrinsic social meaning' (1995: 124). As identification scientists we observe and describe the body, we categorise and individualise, we attempt to answer the simple question 'who are you'? But the bodies, or fragments, reported in the scientific literature are disconnected from the lived world. Social forces produce messy variables that in practice need to be accounted for. Recently, this messiness is being harnessed; a new body is emerging for those studying human identity and identification, one forged by society and layered by time.

References

Abu El-Haj, N. (2007). The genetic reinscription of race. *Annual Review of Anthropology* **36**: 283–300.

Agarwal, S. C. & Beauchesne, P. (2011). It is not carved in bone. Development and plasticity of the aged skeleton. In S. C. Agarwal & B. A. Glencross, eds. *Social Bioarchaeology*. Oxford: Wiley-Blackwell, pp. 312–32.

Agarwal, S. C. & Glencross, B. A. (2011). *Social Bioarchaeology*. Oxford: Wiley-Blackwell.

Ahmed, S. (2002). Racialised bodies. In M. Evans & E. Lee, eds. *Real Bodies: A Sociological Introduction*. New York: Palgrave, pp. 46–63.

Ahmed, S. & Stacey, J. (2001). Introduction: Dermographies. In S. Ahmed & J. Stacey, eds. *Thinking Through the Skin*. London: Routledge, pp. 1–17.

Aiello, L. C. & Molleson, T. (1993). Are microscopic ageing techniques any more accurate than macroscopic ageing techniques? *Journal of Archaeological Science*, **20**: 689–704.

Akridge, M., Hilgers, K. K., Silveira, A. M. et al. (2007). Childhood obesity and skeletal maturation assessed with Fishman's hand-wrist analysis. *American Journal of Orthodontics and Dentofacial Orthopedics*, **132**: 185–90.

Alaeddini, R., Walsh, S. J. & Abbas, A. (2010). Forensic implications of genetic analyses from degraded DNA: a review. *Forensic Science International: Genetics*, **4**: 148–57.

Alberink, I. B., Ruifrok, A. C. C. & Kieckhoefer, H. (2006). Interoperator test for anatomical annotation of earprints. *Journal of Forensic Sciences*, **51**: 1246–54.

Allbrook, D. (1961). The estimation of stature in British and East African males. *Journal of Forensic Medicine*, **8**: 15–28.

Allen, G. E. (1997). The social and economic origins of genetic determinism: a case history of the American Eugenics Movement, 1900–1940, and its lessons for today. *Genetica*, **99**: 77–88.

Al-Raisi, A. N. & Al-Khouri, A. M. (2008). Iris recognition and the challenge of homeland and border control security in UAE. *Telematics and Informatics*, **25**: 117–32.

Alunni-Perret, V., Muller-Bolla, M., Laugier, J.-P. et al. (2005). Scanning electron microscopy analysis of experimental bone hacking trauma. *Journal of Forensic Science*, **50**: 1–6.

Ambrose, S. H., Buikstra, J. & Krueger, H. W. (2003). Status and gender differences in diet at Mound 72, Cahokia, revealed by isotopic analysis of bone. *Journal of Anthropological Archaeology*, **22**: 217–26.

Andrushko, V. A. & Verano, J. W. (2008). Prehistoric trepanation in the Cuzco region of Peru: a view into an ancient Andean practice. *American Journal of Physical Anthropology*, **137**: 4–13.

Angrosino, M. V. (1994). Comments. *Current Anthropology*, **35**: 242–3.

Anidjar, G. (2005). Lines of blood: *limpieza de sangre* as political theology. In M. G. Bondio, ed. *Blood in History and Blood Histories*. Florence: Sismel.

Aoki, K. (2002). Sexual selection as a cause of human skin colour variation: Darwin's hypothesis revisited. *Annals of Human Biology*, **29**: 589–608.

Août, K. B., Pataky, T. C., Clercq, D. & Aerts, P. (2009). The effects of habitual footwear use: foot shape and function in barefoot walkers. *Footwear Science*, **1**: 81–94.

Arber, S. & Ginn, J. (1991). *Gender and Later Life*. London: Sage.

Arber, S. & Ginn, J. (1995). Choice and constraint in the retirement of older married women. In S. Arber & J. Ginn, eds. *Connecting Gender and Ageing: A Sociological Approach*. Buckingham: Philadelphia Press, pp. 69–86.

Arcini, C. (2006). The Vikings bared their filed teeth. *American Journal of Physical Anthropology*, **128**: 727–33.

Armelagos, G. J. (1994). Racism and physical anthropology: Brue's review of Barkan's 'The Retreat of Scientific Racism'. *American Journal of Physical Anthropology*, **93**: 381–3.

Ashbaugh, D. R. (1999). *Quantitative-Qualitative Friction Ridge Analysis: An Introduction to Basic and Advanced Ridgeology*. Boca Raton: CRC Press.

Ashikari, M. (2005). Cultivating Japanese whiteness: the whitening cosmetics boom and the Japanese identity. *Journal of Material Culture*, **10**: 73–91.

Atkinson, P., Glasner, P. & Lock, M. (2009). Genetics and society: perspectives from the twenty-first century. In P. Atkinson, P. Glasner & M. Lock, eds. *Handbook of Genetics and Society. Mapping the New Genomic Era*. Oxford: Routledge, pp. 1–14.

Aufderheide, A. C. (2011). Soft tissue taphonomy: a paleopathology perspective. *International Journal of Paleopathology*, **1**: 75–80.

Aufderheide, A. C. & Rodríguez-Martín, C. (1998). *The Cambridge Encyclopaedia of Human Palaeopathology*. Cambridge University Press.

Averett, S. L., Sikora, A. & Argys, L. M. (2008). For better or worse: relationship status and body mass index. *Economics and Human Biology*, **6**: 330–49.

Ayer, A., Campbell, A., Appelboom, G. et al. (2010). The sociopolitical history and physiological underpinnings of skull deformation. *Neurosurgical Focus*, **29**, DOI: 10.3171/2010.9.

Babler, W. J. (1987). Prenatal development of dermatoglyphic digital patterns: associations with epidermal ridge, volar pad and bone morphology. *Collegium Antropologicum*, **11**: 297–303.

Back, L. (2004). Inscriptions of love. In H. Thomas & J. Ahmed, eds. *Cultural Bodies. Ethnography and Theory.* Oxford: Blackwell, pp. 27–54.

Bahn, P. G. (1987). Getting blood from stone tools. *Nature*, **350**: 14.

Baker, B. J., Dupras, T. L. & Tocheri, M. W. (2005). *The Osteology of Infants and Children.* Texas: A&M University Press.

Bardell, D. (1978). William Harvey 1578–1657. Discovery of the circulation of blood: in commemoration of the 400th anniversary of his birth. *Bioscience*, **28**: 257–9.

Barnes, I. & Thomas, M. G. (2006). Evaluating bacterial pathogen DNA preservation in museum osteological collections. *Proceedings of Biological Science*, **273**: 645–53.

Barry, D. & Petry, N. (2008). Gender differences in associations between stressful life events and body mass index. *Preventative Medicine*, **47**: 498–503.

Barry, D. & Petry, N. M. (2009). Associations between body mass index and substance use disorders differ by gender: results from the National Epidemiologic Survey on Alcohol and Related Conditions. *Addictive Behaviours*, **34**: 51–60.

Bartelink, E. J., Wiersema, J. M., & Demaree, R. S. (2001). Quantitative analysis of sharp-force trauma: an application of scanning electron microscopy in forensic anthropology. *Journal of Forensic Science*, **46**: 1288–93.

Beall, C. M. (1984). Theoretical dimensions of a focus on age in physical anthropology. In D. I. Kertzer & J. Keith, eds. *Age and Anthropological Theory.* Ithaca: Cornell University Press, pp. 82–95.

Beaumont, J., Gledhill, A. & Lee-Thorp, J. (2012). Childhood diet: a closer examination of the evidence from dental tissues using stable isotope analysis of incremental human dentine. *Archaeometry.* doi:10.1111/j.1475-4754.2012.00682.x.

Bennett, J. L. & Benedix, D. C. (1999). Positive identification of remains recovered from an automobile based on presence of an internal fixation device. *Journal of Forensic Sciences*, **44**: 1296–8.

Benthien, C. (2002). *Skin. On the Cultural Border Between Self and the World.* New York: Columbia University Press.

Bentley, R. A. (2006). Strontium isotopes from the Earth to the archaeological skeleton: a review. *Journal of Archaeological Method and Theory*, **13**: 135–87.

Bentley, R. A., Pietrusewsky, M., Douglas, M. T. & Atkinson, T. C. (2005). Matrilocality during the prehistoric transition to agriculture in Thailand? *Antiquity*, **79**: 63–6.

Bentley, R. A., Price, T. D., Lüning, J. et al.(2002). Human migration in early Neolithic Europe. *Current Anthropology*, **43**: 799–804.

Bersaglieri, T., Sabeti, P. C., Patterson, N. et al. (2004). Genetic signatures of strong recent positive selection at the Lactase gene. *American Journal of Human Genetics*, **74**: 1111–20.

Bertillon, A. (1885). La couleur de l'iris. *Revue Scientifique*, **3**: 65–73.

Bevel, T. & Gardner, R. M. (2008). *Bloodstain Pattern Analysis.* 3rd edn. Boca Raton: CRC Press, Inc.

Bianchi, L. & Liò, P. (2007). Forensic DNA and bioinformatics. *Briefings in Bioinformatics*, **8**: 117–28.

Bianucci, R. G., Mattutino, R. & Lallo, P. et al. (2008). Immunological evidence of *Plasmodium falciparum* infection in an Egyptian child mummy from the Early Dynastic Period. *Journal of Archaeological Science*, **35**: 1880–5.

Bildhauer, B. (2006). *Medieval Blood*. Cardiff: University of Wales Press.

Billinger, M. (2007). Another look at ethnicity as a biological concept: moving anthropology beyond the race concept. *Critique of Anthropology*, **27**: 5–35.

Bischoff, J. E., Arrunda, E. M. & Grosh, K. (2000). Finite element modelling of human skin using an isotropic, nonlinear elastic constitutive model. *Journal of Biomechanics*, **33**: 645–52.

Black, S. M. & Thompson, T. J. U. (2007). Body modifications. In T. J. U. Thompson & S. M. Black, eds. *Forensic Human Identification: An Introduction*. Boca Raton: CRC Press.

Blakey, M. J. (1987). Skull doctors: intrinsic social and political bias in the history of physical anthropology with special reference to the work of Ales Hrdlicka. *Critique of Anthropology*, **7**: 7–35.

Blakey, M. L. (2001). Bioarchaeology of the African diaspora in the Americas: its origins and scope. *Annual Review of Anthropology*, **30**: 387–422.

Blane, D. (2006). The life course, the social gradients, and health. In M. Marmot & R. G. Wilkinson, eds. *Social Determinants of Health*. 2nd edn. Oxford University Press, pp. 54–77.

Blom, D. (2005). Embodying borders: human body modification and diversity in Tiwanaku society. *Journal of Anthropological Archaeology*, **24**: 1–24.

Boas, F. (1912). Changes in the bodily form of descendents of immigrants. *American Anthropology*, **14**: 530–62.

Bock, J. & Sellen, D. W. (2002). Introduction to a special issue on childhood and the evolution of the human life course. *Human Nature*, **15**: 63–81.

Bocquet-Appel, J.-P. & Masset, C. (1982). Farewell to palaeodemography. *Journal of Human Evolution*, **11**: 321–33.

Bodziak, W. J. (2000). *Footwear Impression Evidence: Detection, Recovery, and Examination*. 2nd edn. Boca Raton: CRC Press.

Boechat, M. S., Pietka, E., Huang, H. K. & Gilsanz, V. (2001). Skeletal age determinations in children of European and African decent: applicability of the Greulich Pyle standards. *Pediatric Research*, **50**: 624–8.

Boellstorff, T. (2008). *Coming of Age in Second Life: An Anthropologist Explores the Virtual Human*. New Jersey: Princeton University Press.

Bogin, B. (2001). *The Growth of Humanity*. New York: Wiley-Liss.

Bogin, B. (2005). Patterns of Human Growth. Cambridge University Press.

Bordo, S. (1993). *Unbearable Weight: Feminism, Western Culture and the Body*. Berkeley: University of California Press.

Borić, D, & Robb, J., eds. (2008). *Past Bodies: Body Centred Research in Archaeology*. Oxford: Oxbow.

Bouwman A. S., & Brown, T. A. (2005). The limits of biomolecular palaeopathology: ancient DNA cannot be used to study venereal syphilis. *Journal of Archaeological Science*, **32**: 691–702.

Boylan, M. (2007). Galen: on blood, the pulse and the arteries. *Journal of the History of Biology*, **40**: 207–36.

Boylston, A., Fiorato, V. & Knüsel, C. J. (2000). *Blood Red Roses: The Archaeology of a Mass Grave from the Battle of Towton AD 1461*. Oxford: Oxbow.

Brace, L. C. (1995). Region does not mean 'race' – reality versus convention in forensic anthropology. *Journal of Forensic Sciences*, **40**: 171–5.

Bradley, H. (1996). *Fractured Identities*. Cambridge: Polity Press.

Bradley, E. J., Nabhani, F. & Spears, I. R. (2009). The effect of the degree of screw tension on interfragmentary displacement in stabilized fractures of the femoral neck. *Current Orthopaedic Practice*, **20**: 291–9.

Braithwaite, R., Robillard, A., Woodring, T., Stephens, T. & Arriola, K. J. (2001). Tattooing and body piercing amongst adolescent detainees: relationship to alcohol and other drug use. *Journal of Substance Abuse*, **13**: 5–16.

Branchet, M. C., Boisnic, S., Frances, C. & Robert, A. M. (1990). Skin thickness changes in normal aging skin. *Gerontology*, **36**: 28–35.

Brandt, E., Weichmann, I. & Grupe, G. (2002). How reliable are immunological tools for the detection of ancient protein in fossil bones? *International Journal of Osteoarcheology*, **12**: 307–16.

Braun, B. (2007). Biopolitics and the molecularization of life. *Cultural Geographies*, **14**: 6–28.

Breeze, R. G., Budowle, B. & Schutzer, S. E., eds. (2005). *Microbial Forensics*. New York: Elsevier Academic Press.

Brickley, M., Buteux, S., Adams, J. & Cherrington, R., eds. (2006). *St Martin's Uncovered. Investigations in the Churchyard of St Martin's-in-the-Bull Ring, Birmingham, 2001*. Oxford: Oxbow.

Brickley, M. & Ives, R. (2008). *The Bioarchaeology of Metabolic Bone Disease*. London: Academic Press.

Brickley, M. & McKinley, J. I., eds. (2004). *Guidance to the Standards for Recording Human Skeletal Remains*. Reading: Institute of Field Archaeologists/British Association of Biological Anthropology and Osteoarchaeology.

Brock, D. W. (2002). Human cloning and our sense of self. *Science*, **296**: 314–16.

Broeders, A. P. A. (2006). Of earprints, fingerprints, scent dogs, cot deaths and cognitive contamination – a brief look at the present state of play in the forensic arena. *Forensic Science International*, **159**: 148–57.

Brooks, T. L., Woods, E. R., Knight, J. R. & Shrier, L. A. (2003). Body modification and substance use in adolescents: is there a link? *Journal of Adolescent Health*, **32**: 44–9.

Brown, T. A. (2001). Ancient DNA. In D. R. Brothwell & A. M. Pollard, eds. *Handbook of Archaeological Sciences*. Chichester: John Wiley and Sons, Ltd.

Brown, T. & Brown, K. (2011). *Biomolecular Archaeology: An Introduction*. Oxford: Wiley-Blackwell.

Brown, K. A., Elliot, T. R., Rogers, A. H. & Thonard, J. C. (1984). The survival of oral *streptococci* on human skin and its implications in bite-mark investigation. *Forensic Science International*, **26**: 193–7.

Brown, N. & Webster, A. (2004). *New Medical Technologies and Society*. London: Polity Press.

Bruce, V., Green, P. R. & Georgeson, M. A. (2003). *Visual perception: physiology, psychology and ecology*. 4th edn. Sussex: Psychology Press.

Brunner, E. & Marmot, M. (2006). Social organization, stress and health. In M. Marmot & R. G. Wilkinson, eds. *Social Determinants of Health*. 2nd edn. Oxford University Press, pp. 7–30.

Budd, P., Millard, A., Chenery, C., Lucy, S. & Roberts, C. (2004). Investigating population movement by stable isotope analysis: a report from Britain. *Antiquity*, **78**: 127–41.

Budgeon, S. (2003). Identity as an embodied event. *Body and Society*, **9**: 35–55.

Buikstra, J. E. & Milner, G. R. (1991). Isotopic and archaeological interpretations of diet in the Central Mississippi Valley. *Journal of Archaeological Science* **18**: 319–29.

Buikstra, J. & Ubelaker, D. H., eds. (1994). *Standards for Data Collection from Human Skeletal Remains*. Proceedings of a seminar at the Field Museum of Natural History organized by Jonathan Haas. Arkansas: Arkansas Archaeological Survey Research Series No. 44.

Burge, M. & Burger, W. (1999). Ear biometrics. In A. K. Jain, R. Bolle & S. Pankanti, eds. *Biometrics: Personal Identification in Networked Society*. USA: Kluwer Academic Publishers, pp. 274–85.

Burnett, J. (1976). English diet in the eighteenth and nineteenth centuries. *Progress in Food and Nutrition Science*, **2**: 11–34.

Bury, M. (1995). Ageing, gender and sociological theory. In S. Arber & J. Ginn, eds. *Connecting Gender and Ageing: a Sociological Approach*. Buckingham: Open University Press, pp. 15–29.

Butler, J. (1990). *Gender Trouble: Feminism and the Subversion of Identity*. New York: Routledge.

Butler, J. (1993). *Bodies that Matter: On the Discursive Limits of 'Sex'*. New York: Routledge.

Calleja-Agius, J., Muscat-Baron, Y. & Brincat, M. P. (2007). Skin ageing. *Menopause International*, **13**: 60–4.

Campbell, R. T. & Alwin, D. F. (1996). Quantitative approaches: toward an integrated science of aging and human development. In R. H. Binstock & L. K. George, eds. *Handbook of Aging and the Social Sciences*. 4th edn. New York: Academic Press, pp. 31–51.

Cann, R. L., Stoneking, M. & Wilson, A. C. (1987). Mitochondrial DNA and human evolution. *Nature*, **325**: 31–6.

Capelli, C., Tschentscher, F. & Pascali, V. L. (2003). 'Ancient' protocols for the crime scene? Similarities and differences between forensic genetics and ancient DNA analysis. *Forensic Science International*, **131**: 59–64.

Cardoso, H. F. V. (2010). Testing discriminant functions for sex determination from deciduous teeth. *Journal of Forensic Sciences*, **55**: 1557–60.

Carey, H. M. (2010). Judicial astrology in theory and practice in later medieval Europe. *Studies in History and Philosophy of Biological and Biomedical Sciences*, **41**: 90–8.

Carlino, A. (1999). *Books of the Body*. University of Chicago Press.

Carsten, J. (2011). Substance and relationality: blood in contexts. *Annual Review of Anthropology*, **40**: 19–35.

Cartmill, M. (1979). The volar skin of primates: its frictional characteristics and their functional significance. *American Journal of Physical Anthropology*, **50**: 497–509.

Caspari, R. (2009). 1918: Three perspectives on race and human variation. *American Journal of Physical Anthropology*, **139**: 5–15.

Castelo-Branco, C., Pons, F., Gratacós, E. et al. (1994). Relationship between skin collagen and bone changes during aging. *Maturitas*, **18**: 199–206.

Cattaneo, C. (2007). Forensic anthropology: developments of a classical discipline in the new millennium. *Forensic Science International*, **165**: 185–93.

Cattaneo, C., Gelsthorpe, K., Phillips, P. & Sokol, R. J. (1990). Blood in ancient human bones. *Nature*, **347**: 339.

Cattaneo, C., Gelsthorpe, K., Phillips, P. & Sokol, R. J. (1995). Differential survival of albumin in ancient bone. *Journal of Archaeological Science*, **22**: 271–6.

Cattaneo, C., Gelsthorpe, K., Sokol, R. J. & Phillips, P. (1994). Immunological detection of albumin in ancient human cremations using ELISA and monoclonal antibodies. *Journal of Archaeological Science*, **21**: 565–71.

Cauna, N. (1954). Nature and functions of the papillary ridges of the digital skin. *Anatomical Review*, **119**: 449–68.

Chadwick, E. (1965) *The Sanitary Conditions of the Labouring Population of Great Britain: Report 1842*. Edinburgh University Press.

Chamberlain, A. (2006). *Demography in Archaeology*. Cambridge University Press.

Chambers, S. A. (2007). Sex and the problem of the body: reconstructing Judith Butler's theory of sex/gender. *Body and Society*, **13**: 47–75.

Champod, C., Lennard, C., Margot, P. & Stoilovic, M. (2004). *Fingerprints and Other Ridge Skin Impressions*. Boca Raton: CRC Press.

Chandramohan, A. & Sivasankar, V. (2009). Role of rag-pickers in the management of solid waste in Tiruchirappalli City, Tamil Nadu, India. *Waste Management*, **29**: 3052–4.

Chaplin, G. & Jablonski, N. G. (2009). Vitamin D deficiency and the evolution of human depigmentation. *American Journal of Physical Anthropology*, **139**: 451–61.

Chapman, C. L. (1993). Alphonse M. Bertillon: his life and the science of fingerprints. *Journal of Forensic Identification*, **43**: 585–603.

Chapman, J. (2010). 'Deviant' burials in the Neolithic and Chalcolithic of Central and South Eastern Europe. In K. Rebay-Salisbury, M. L. Stig Sørensen & J. Hughes, eds. *Body Parts and Body Whole*. Oxford: Oxbow, pp. 30–45.

Chaudhry, R. & Pant, S. K. (2004). Identification of authorship using lateral palm print – a new concept. *Forensic Science International*, **141**: 49–57.

Chen, E., Miller, G. E., Walker, H. A. et al. (2008). Genome-wide transcriptional profiling linked to social class in asthma. *Thorax*, **64**: 38–43.

Chen, J., Moon, Y-S., Wong, M-F. & Su, G. (2010). Palmprint authentication using a symbolic representation of images. *Image and Vision Computing*, **28**: 343–51.

Chenery, C., Müldner, G., Evans, J., Eckardt, H., & Lewis, M. (2010). Strontium and stable isotope evidence for diet and mobility in Roman Gloucester, UK. *Journal of Archaeological Science*, **37**: 150–63.

Cherryson, A. (2010). In the pursuit of knowledge: Dissection, post-mortem surgery and the retention of body parts on 18th and 19th century Britain. In K. Rebay-Salisbury, M. L. Stig Sørensen & J. Hughes, eds. *Body Parts and Body Whole*. Oxford: Oxbow, pp. 135–48.

Christensen, A. M. & Crowder, C. M. (2009). Evidentiary standards for forensic anthropology. *Journal of Forensic Sciences*, **54**: 1211–16.

Clark, H. R., Goyder, E., Bissell, P. et al. (2007). A pilot survey of socio-economic differences in child-feeding behaviours among parents of primary-school children. *Public Health Nutrition*, **11**: 1030–6.

Clarke, A. E., Shim, J., Shostak, S., & Nelson, A. (2009). Biomedicalising genetic health, diseases and identities. In P. Atkinson, P. Glasner & M. Lock, eds. *Handbook of Genetics and Society. Mapping the New Genomic Era*. Oxford: Routledge, pp. 21–40.

Clarkson, J. & Schaefer, M. (2007). Surgical intervention. In T. J. U. Thompson & S. M. Black, eds. *Forensic Human Identification: An Introduction*. Boca Raton: CRC Press, pp. 127–46.

Cocilovo, J. A., Varela, H. H. & O'Brien, T. G. (2011). Effects of artificial deformation on cranial morphogenesis in the south central Andes. *International Journal of Osteoarchaeology*, **21**: 300–12.

Cole, S. A. (2000). The myth of fingerprints: a forensic science stands trial. *Lingua Franca*, **10**: 54–62.

Cole, S. W. (2009) The social regulation of human gene expression. *Current Directions in Psychological Science*, **18**(3): 132–7.

Coles, C. (1990). The older women in Hausa society. In J. Soklovsky, ed. *The Cultural Context of Ageing*. London: Bergen and Garvey, pp. 57–81.

Comstock, R. D., Castillo, E. M. & Lindsay, S. P. (2004). Four-year review of the use of race and ethnicity in epidemiologic and public health research. *American Journal of Epidemiology*, **159**: 611–19.

Connor, S. (2001). Mortification. In S. Ahmed & J. Stacey, eds. *Thinking Through the Skin*. New York: Routledge, pp. 36–51.

Connor, S. (2004). *The Book of Skin*. Ithaca: Cornell University Press.

Cook, B. & Zelhof, A. C. (2008). Photoreceptors in evolution and disease. *Nature Genetics*, **40**: 1275–6.

Copeman, J. (2009). Introduction: blood donation, bioeconomy, culture. *Body and Society* **15**: 1–28.

Coughlin, J. W. & Guarda, A. S. (2006). Behavioural disorders affecting food intake: eating disorders and other psychiatric conditions. In M. E. Shils, M. Shike, A. C. Ross, B. Caballero & R. J. Cousins, eds. *Modern Nutrition in Health and Disease*. 10th edn. Baltimore: Lippincott Williams & Wilkins, pp. 1353–61.

Craig, E. F. & Buckberry, J. L. (2010). Investigating social status using evidence of biological status: a case study from Raunds Furnells. In. J. Buckberry & A. Cherryson, eds. *Burial in Later Anglo-Saxon England c650–1100 AD*. Oxford: Oxbow Books, pp. 128–42.

Crawford, P. (2004). *Blood, Bodies and Families in Early Modern England*. Edinburgh: Pearson Longman.

Crawford, S. (1991). When do Anglo-Saxon children count? *Journal of Theoretical Archaeology*, **2**: 17–24.

Crawford, S. (2010). Differentiation in the later Anglo-Saxon burial ritual on the basis of mental or physical impairment: a documentary perspective. In. J. Buckberry & A. Cherryson, eds. *Burial in Later Anglo-Saxon England c650–1100 AD*. Oxford: Oxbow Books, pp. 93–102.

Crawshaw, P. (2007). Governing the healthy male citizen: men, masculinity and popular health in men's health magazines. *Social Science & Medicine*, **65**: 1606–18.

Crawshaw, P. (2009). Critical perspectives on the health of men: lessons from medical sociology. *Critical Public Health*, **19**: 279–85.

Cream, J. (1994). 'Out of place' paper presented at the annual meeting of the Association of American Geographers, San Francisco, Mar-Apr.

Crews, D. E. (2003). *Human Senescence. Evolutionary and Biocultural Perspectives*. Cambridge University Press.

Crossland, Z. (2002). Violent spaces: conflict over the reappearance of Argentina's disappeared. In J. Schofield, C. Beck, & W. G. Johnson, eds. *The Archaeology of 20th Century Conflict*. Routledge: London, pp. 115–31.

Crossland, Z. (2009a). Acts of estrangement: the making of self and other through exhumation. *Archaeological Dialogues* **16**(1): 102–25.

Crossland, Z. (2009b). Of clues and signs: the dead body and its evidential traces. *American Anthropologist*, **111**(1): 69–80.

Crow, F., Sperduti, A., O'Connell, T. et al. (2010). Water-related occupation and diet in two Roman coastal communities (Italy, first to third century AD): correlation between stable carbon and nitrogen isotope values and auricular exostosis prevalence. *American Journal of Physical Anthropology*, **142**: 355–66.

Cummings, S. R., Ling, X. & Stone, K. (1997). Consequences of foot binding among older women in Beijing, China. *American Journal of Public Health*, **87**: 1677–9.

Dao, H. & Kazin, R. A. (2007). Gender differences in the skin: a review. *Gender and Medicine*, **4**: 308–28.

Darmon, N. & Drewnowski, A. (2008). Does social class predict diet quality? *American Journal of Clinical Nutrition*, **87**: 1107–17.

Daughman, J. (1999). Recognizing persons by their iris patterns. In A. K. Jain, R. Bolle & S. Pankanti, eds. *Biometrics: Personal Identification in Networked Society*. New York: Kluwer Academic Publishers, pp. 103–21.

Daughman, J. (2003). The importance of being random: statistical principles of iris recognition. *Pattern Recognition*, **36**: 279–91.

Daughman, J. (2007). Identifying persons from their iris patterns. In T. J. U. Thompson & S. M. Black, eds. *Forensic Human Identification: An Introduction*. Boca Raton: CRC Press, pp. 271–85.

De Beauvoir, S. (1974). *The Second Sex.* (trans and ed by H. M. Parshley). New York: Vintage Books.

De Souza Ferreira, J. E. & da Veiga, G. V. (2008). Eating disorder risk behaviour in Brazilian adolescents from low socio-economic level. *Appetite*, **51**: 249–55.

De Valck, E. (2006). Major incident response: collecting ante-mortem data. *Forensic Science International*, **159**S: S15–S19.

Delalleau, A., Josse, G., Lagarde, J. M., Zahuani, H. & Bergheau, J. M. (2008). Characterization of the mechanical properties of skin by inverse analysis combined with an extensometry test. *Wear*, **264**: 405–10.

Delphy, C. (1993). Rethinking sex and gender. *Women's Studies International Forum*, **16**(1): 1–9.

Dempsey, E. (2005). *The Absorption of Tattoo Dye into the Lymphatic System.* Unpublished dissertation, University of Dundee.

Díaz-Andreu, M. & Lucy, S. (2005). Introduction. In M. Díaz-Andreu, S. Lucy, S. Babić & D. N. Edwards, eds. *The Archaeology of Identity: Approaches to Gender, Age, Status, Ethnicity and Religion.* London: Routledge, pp. 1–12.

Dibdin, M. (1992). The pathology lesson. *Granta*, **39**: 91–101.

Dietz, W. H. (2006). Childhood obesity. In M. E. Shils, M. Shike, A. C. Ross, B. Caballero & R. J. Cousins, eds. *Modern Nutrition in Health and Disease.* 10th edn. Baltimore: Lippincott Williams & Wilkins, pp. 979–90.

Diridollou, S., Berson, M. & Vabre, V., eds. (1998). An *in vivo* method for measuring the mechanical properties of the skin using ultrasound. *Ultrasound in Medicine and Biology*, **24**: 215–24.

Donaldson, A. E., Taylor, M. C., Cordiner, S. J. & Lamont, I. L. (2010). Using oral microbial DNA analysis to identify expirated bloodspatter. *International Journal of Legal Medicine*, **124**: 569–76.

Dorling, D., Shaw, M. & Brimblecombe, N. (2000). Housing wealth and community health: exploring the role of migration. In H. Graham, ed. *Understanding Health Inequalities.* Buckingham: Open University Press, pp. 186–202.

Dowling, A. M., Steele, J. R. & Baur, L. A. (2001). Does obesity influence foot structure and plantar pressure patterns in prepubescent children? *International Journal of Obesity*, **25**: 845–52.

Dressler, W. W., Oths, K. S. & Gravlee, C. C. (2005). Race and ethnicity in public health research: models to explain health disparities. *Annual Review of Anthropology*, **34**: 231–52.

Easterbrook, G. (2001). A moral precedent for the Siamese twins case. *Human Life Review*, **27**: 37–40.

Echarri, J. J. & Forriol, F. (2003). The development in footprint morphology in 1851 Congolese children from urban and rural areas, and the relationship between this and wearing shoes. *Journal of Pediatric Orthopaedics B*, **12**: 141–6.

Edgar, H. J. H. & Hunley, K. L. (2009). Race reconciled?: How biological anthropologists view human variation. *American Journal of Physical Anthropology*, **139**: 1–4.

Elliott, J. (2011). Passport to payment. *Biometric Technology Today*, **6**: 5–8.

Elliott, M. & Collard, M. (2009). FORDISC and the determination of ancestry from cranial measurements. *Biology Letters* **5**: 849–52.

Emler, N. (2005). *Life Course Transitions and Social Identity Change.* In P. Levy, P. Ghisetta, J.-M. Le Goff, D. Spini, eds. *Towards an Interdisciplinary Perspective on the Life Course,* New York: Elsevier, pp. 203–21.

Endicott, P. & Ho, S. Y. W. (2008). A Bayesian evaluation of human mitochondrial substitution rates. *The American Journal of Human Genetics,* **82**: 895–902.

Engels, F. (1987 [1845]). *The Conditions of the Working Class in England.* Harmondsworth: Penguin.

Enns, J. T. (2004). *The Thinking Eye, the Seeing Brain. Explorations in Visual Cognition.* New York: Norton.

Entwhistle, J. (2002). The dressed body. In M. Evans & E. Lee, eds. *Real Bodies: A Sociological Introduction.* New York: Palgrave, pp. 133–50.

Epstein, C. J. (2003). Is modern genetics the new eugenics? *Genetics in Medicine,* **5**: 469–75.

Epstein, R. (2002). Fingerprints meet *Daubert:* The myth of fingerprint 'science' is revealed. *Southern Californian Law Review,* **75**: 605–55.

Epstein, S. (2004). Bodily differences and collective identities: the politics of gender and race in biomedical research in the United States. *Body and Society,* **10**: 183–203.

Escoffier, C., de Rigal, J., Rochefort, A. et al. (1989). Age-related mechanical properties of human skin: an *in vivo* study. *Journal of Investigative Dermatology,* **93**: 353–7.

Evans, M. (2002). Real bodies: an introduction. In M. Evans & E. Lee, eds. *Real Bodies: A Sociological Introduction.* New York: Palgrave, pp. 1–13.

Evans, J. H. & Schairer, C. (2009). Bioethics and human genetic engineering. In P. Atkinson, P. Glasner & M. Lock, eds. *The Handbook of Genetics and Society: Mapping the New Genomic Era.* London: Routledge, pp. 349–66.

Falanga, V. & Bucalo, B. (1993). Use of a durometer to assess skin hardness. *Journal of the American Academy of Dermatology,* **29**: 47–51.

Falys, C. G., Schutkowski, H. & Weston, D. A. (2006). Auricular surface aging: Worse than expected? A test of the revised method on a documented historic skeletal assemblage. *American Journal of Physical Anthropology,* **130**: 508–13.

Farah, M. J. (2004). *Visual Agnosia.* 2nd edn. Cambridge: MIT Press.

Fausto-Sterling, A. (2005). The bare bones of sex: part 1. Sex and gender. *Signs,* **30**: 1491–1527.

Featherstone, M. (1999). Body modification: an introduction. *Body and Society,* **5**: 1–13.

Feldmeier, H. & Heukelbach, J. (2009). Epidermal parasitic skin diseases: a neglected category of poverty-associated plagues. *Bulletin of the World Health Organisation,* **87**: 152–9.

Ferenc, M. (2008). Anatomical considerations in bloodstain pattern analysis. In T. Bevel & R. M. Gardner, eds. *Bloodstain Pattern Analysis.* 3rd edn. Boca Raton: CRC Press, pp. 135–48.

Fieldhouse, S. (2011). Consistency and reproducibility in fingermark deposition. *Forensic Science International*, **207**: 96–100.

Fields, R. & Molina, D. K. (2008). A novel approach for fingerprinting mummified hands. *Journal of Forensic Sciences*, **53**: 952–5.

Fildes, V. A. (1988). *Wet-Nursing: A History from Antiquity to the Present*. Oxford: Blackwell.

Fiorato, V., Boylston, A. & Knüsel, C. (2000). *Blood Red Roses: The Archaeology of a Mass Grave from the Battle of Towton AD1461*. Oxford: Oxbow.

Fisher, G. J., Varani, J. & Voorhees, J. J. (2008). Looking older: fibroblast collapse and therapeutic implications. *Archives of Dermatology*, **144**: 666–72.

Fisher, N. (2008). It's not what you know, it's who you are! *Biometric Technology Today* **16**: 7–9.

Flohr, C., Nguyen Tuyen, L., Lewis, S. et al. (2006). Poor sanitation and helminth infection protect against skin sensitization in Vietnamese children: a cross-sectional study. *Journal of Allergy and Clinical Immunology*, **118**: 1305–11.

Flöistrup, H., Swartz, J., Bergstrom, A. et al. (2006). Allergic disease and sensitization in Steiner school children. *Journal of Allergy and Clinical Immunology*, **117**: 59–66.

Fornaciari, G., Giuffra, V. & Ferroglio, E. (2010). *Plasmodium falciparum* immunodetection in bone remains of members of the Renaissance Medici family (Florence, Italy, sixteenth century). *Transactions of the Royal Society of Tropical Medicine and Hygiene*, **104**(9): 583–7.

Fortun, M. (2009). Genes in our knot. In P. Atkinson, P. Glasner & M. Lock, eds. *Handbook of Genetics and Society. Mapping the New Genomic Era*. Oxford: Routledge, pp. 247–59.

Fortes, M. (1984). Age, generation, and social structure. In D. I. Kertzer & J. Keith, eds. *Age and Anthropological Theory*. Ithaca: Cornell University Press, pp. 99–122.

Fournier, V. (2002). Fleshing out gender: crafting gender identity on women's bodies. *Body and Society*, **8**: 55–77.

Frank, A. (1991). For a sociology of the body, an analytical review. In M. Featherstone, M. Hepworth & B. Turner, eds. *The Body, Social Process and Cultural Theory*. London: Sage, pp. 36–102.

Frey, C., Thompson, F., Smith, J., Sanders, M. & Horstman, H. (1993). American orthopaedic foot and ankle society women's shoe survey. *Foot and Ankle* **14**: 78–81.

Frost, H. M. (2004). A 2003 update on bone physiology and Wolff's Law for clinicians. *Angle Orthodontist*, **74**: 3–15.

Frumkin, D., Wasserstrom, A., Davidson, A. & Grafit, A. (2010). Authentication of forensic DNA samples. *Forensic Science International: Genetics*, **4**: 95–103.

Fry, C. (1996). Age, aging and culture. In R. H. Binstock & L. K. George, eds. *Handbook of Aging and the Social Sciences*. 4th edn. New York: Academic Press, pp. 118–35.

Fuller, B. T., Molleson, T. I., Harris, D. A., Gilmour, L. T. & Hedges, R. E. M. (2006). Isotopic evidence for breastfeeding and possible adult dietary differences from late/sub-Roman Britain. *American Journal of Physical Anthropology*, **129**: 45–54.

Fully, G. (1956). Une nouvelle méthode de détermination de la taille. *Annales de Medecine Legale*, **40**: 145–54.

Gallagher, S. (2000). Philosophical conceptions of the self: implications for cognitive science. *Trends in Cognitive Science*, **4**: 14–21.

Gamble, S. D. (1943). The disappearance of foot-binding in Tinghsien. *The American Journal of Sociology*, **49**(2): 181–3.

Ganeshan, B., Theckedath, D., Young, R. & Chatwin, C. (2006). Biometric iris recognition system using a fast and robust iris localization and alignment procedure. *Optics and Lasers in Engineering*, **44**: 1–24.

Gawkrodger, D. J. (2002). *Dermatology: An Illustrated Colour Text*. 3rd edn. Edinburgh: Churchill Livingstone.

Geiger, H. J. (2006) Health disparities: What do we know? What do we need to know? What should we do? In A. J. Schulz & L. Mullings, eds. *Gender, Race, Class and Health: Intersectional Approaches*, San Francisco: Jossey-Bass, pp. 261–88.

Gell, A. (1993). *Wrapping in Images: Tatooing in Polynesia*. Oxford: Clarendon Press.

Geller, P. L. (2005). Skeletal analysis and theoretical complications. *World Archaeology*, **37**: 597–609.

Geller, P. L. (2006). Altering identities: body modifications and the Pre-Columbian Maya. In R. L. Gowland & C. J. Knüsel, eds. *The Social Archaeology of Funerary Remains*. Oxford: Oxbow, pp. 168–78.

Geller, P. L. (2008). Conceiving sex: fomenting a feminist bioarchaeology. *Journal of Social Archaeology*, **8**: 113–38.

Geller, P. L. (2011). Getting a head start in life: pre-Columbian Maya cranial modification from infancy to ancestorhood. In M. Bonogofsky, ed. *The Bioarchaeology of the Human Head*. University of Florida Press, pp. 241–61.

Giacomoni, P. U., Mammone, T. & Teri, M. (2009). Gender-linked differences in human skin. *Journal of Dermatological Sciences*, **55**: 144–9.

Giddens, A. (1991). *Modernity and Self-Identity. Self and Society in the Late Modern Age*. California: Stanford University Press.

Gilchrist, R. (1999). *Gender and Material Culture: The Archaeology of Religious Women*. London: Routledge.

Gilchrist, R. (2000). Archaeological biographies: realizing human lifecycles, courses, histories. *World Archaeology*, **31**: 325–8.

Gilchrist, R. (2012) *Medieval Life: Archaeology and the Life Course*. Woodbridge: Boydell & Brewer.

Gilleard, C. & Higgs, P. (2000). *Cultures of Ageing: Self Citizen and the Body*. Essex: Prentice Hall, 126–43.

Gillon, R. (2001). Imposed separation of conjoined twins: moral hubris by the English courts? *Journal of Medical Ethics*, **27**: 3–4.

Ginn, J. & Arber, S. (1995). Only connect: gender relations and ageing. In S. Arber & J. Ginn, eds. *Connecting Gender and Ageing: a Sociological Approach*. Milton Keynes: Open University Press, pp. 1–14.

Glendenning, J. A. (2002). Refusal of blood because of religious beliefs: a patient's right to die. *Journal of Emergency Medicine*, **28**: 196–8.

Gonçalves, D., Campanacho, V. & Cardoso, H. F. V. (2011). Reliability of the lateral angle of the internal auditory canal for sex determination of subadult skeletal remains. *Journal of Forensic and Legal Medicine*, **18**: 121–4.

Goodwin, W., Linacre, A. & Hadi, S. (2007). *An Introduction to Forensic Genetics*. London: John Wiley and Sons, Ltd.

Gould, S. J. (1997). *The Mismeasure of Man*. London: Penguin Books.

Gowland, R. L. (2001). Playing dead: implications of mortuary evidence for the social construction of childhood in Roman Britain. In G. Davies, A. Gardner & K. Lockyear, eds. *TRAC 2000. Proceedings of the Tenth Annual Theoretical Roman Archaeology Conference, London 2000*. Oxford: Oxbow, pp. 152–68.

Gowland, R. L. (2002). *Age as an Aspect of Social Identity in Fourth to Sixth Century AD England: The Osteological and Funerary Evidence*. Unpublished PhD Thesis, Durham University.

Gowland, R. L. (2006) Age as an aspect of social identity: the archaeological funerary evidence. In R. L. Gowland & C. J. Knüsel, eds. *The Social Archaeology of Funerary Remains*. Oxford: Oxbow, pp. 143–54.

Gowland, R. L. (2007). Age, ageism and osteological bias: the evidence from late Roman Britain. *Journal of Roman Archaeology Supplementary Series*, **65**: 153–69.

Gowland, R. L. & Chamberlain, A. T. (2002). A Bayesian approach to ageing perinatal skeletal material from archaeological sites: implications for the evidence for infanticide in Roman Britain. *Journal of Archaeological Science*, **29**: 677–85.

Gowland, R. L. & Chamberlain, A. (2005a). Detecting plague: palaeodemographic characterisation of a catastrophic death assemblage. *Antiquity*, **79**: 146–57.

Gowland, R. L. & Chamberlain, A. (2005b). Estimating age-at-death from the pubic symphysis: past, present and future. In M. Clegg & S. Zakrewski, eds. *Proceedings of the British Association for Biological Anthropology and Osteoarchaeology 2003*, Southampton: British Archaeological Reports International Series.

Gowland, R. L. & Knüsel, C. J. (2006). Introduction. *The Social Archaeology of Funerary Remains*. Oxford: Oxbow. In R. L. Gowland & C. J. Knüsel, eds. *The Social Archaeology of Funerary Remains*. Oxford: Oxbow, pp. ix–xiv.

Gowland, R. L. & Redfern, R. C. (2010). Childhood health in the Roman world: perspectives from the centre and margin of the Empire. *Childhood in the Past*, **3**: 15–42.

Graham, E. A. M., Bowyer, V. L., Martin, V. J. & Rutty, G. N. (2008). Investigation into the usefulness of DNA profiling of earprints. *Science & Justice*, **47**: 155–9.

Graham, H. (2000). The challenge of health inequalities. In H. Graham, ed. *Understanding Health Inequalities*. Buckingham: Open University Press, pp. 3–24.

Grauer, A. & Stuart-Macadam, P., eds. (1998). *Sex and Gender in Paleopathological Perspective*. Cambridge University Press.

Gravlee, C. C. (2009). How race becomes biology: embodiment of social inequality. *American Journal of Physical Anthropology*, **139**: 47–57.

Gravlee, C. C., Bernarn, H. K. & Leonard, W. R. (2003). Heredity, environment, and cranial form: a reanalysis of Boas's immigrant data. *American Anthropologist*, **105**: 125–38.

Gravlee, C. C. & Dressler, W. W. (2005). Skin pigmentation, self-perceived color, and arterial blood pressure in Puerto Rico. *American Journal of Human Biology*, **17**: 195–206.

Gravlee, C. C., Dressler, W. W. & Bernard, H. R. (2005). Skin color, social classification and blood pressure in Puerto Rico. *American Journal of Public Health* **95**: 2191–7.

Gray, J. (1993). *Men Are from Mars, Women Are from Venus: a Practical Guide for Improving Communication and Getting What You Want in Your Relationships*. Harper Collins.

Green, R. E., Krause, J., Briggs, A. W. et. al. (2010). A draft sequence of the Neanderthal genome. *Science* **328**, 710–22.

Gregory, R. L. (2007). *Eye and Brain: the Psychology of Seeing*. 5th edn. Oxford University Press.

Grivas, C. R. & Komar, D. A. (2008). *Kumho, Daubert*, and the nature of scientific inquiry: implications for forensic anthropology. *Journal of Forensic Sciences*, **53**: 771–6.

Guadagno, R. E., Muscanell, N. L., Okdie, B. M., Burk, N. M. & Ward, T. B. (2011). Even in virtual environments women shop and men build: a social role perspective on Second Life. *Computers in Human Behaviour* **27**: 304–8.

Guidolin, D., Crivellato, E. & Ribatti, D. (2011). The 'self-similarity logic' applied to the development of the vascular system. *Developmental Biology*, **351**: 156–62.

Gump, W. (2010). Modern induced skull deformity in adults. *Neurosurgical Focus*, **29**: E4.

Gunn, A. (2009). *Essential Forensic Biology*. 2nd edn. London: Wiley-Blackwell.

Haglund, W. & Sorg, M. (1997). *Forensic taphonomy: the postmortem fate of human remains*. Boca Raton: CRC Press.

Halcrow, S. E. & Tayles, N. (2008). The bioarchaeological investigation of childhood and social age: problems and prospects. *Journal of Archaeological Method and Theory*, **15**: 190–215.

Halcrow, S. E. & Tayles, N. (2011). The bioarchaeological investigation of children and childhood. In S. C. Agarwal & B. A. Glencross, eds. *Social Bioarchaeology*. Oxford: Wiley-Blackwell, pp. 333–60.

Halse, C. (2009). Bio-citizenship: virtue discourses and the birth of the bio-citizen. In J. Wright & V. Harwood, eds. *Biopolitics and the Obesity Epidemic*. New York: Routledge, pp. 45–59.

Hamer, M., Witte, D. R., Mosdøl, A., Marmot, M. G. & Brunner, E. J. (2008). Prospective study of coffee and tea consumption in relation to risk of type 2 diabetes mellitus among men and women: the Whitehall II study. *British Journal of Nutrition*, **100**: 1046–53.

Hamrick, M. W. (1998). Functional and adaptive significance of primate pads and claws: evidence from New World Anthropoids. *American Journal of Physical Anthropology*, **106**: 113–27.

Has, C. & Bruckner-Tuderman, L. (2006). Molecular and diagnostic aspects of genetic skin fragility. *Journal of Dermatological Science*, **44**: 129–44.

Hatfield, J. S. (2006). The genetic basis of hair whorl, handedness, and other phenotypes. *Medical Hypotheses*, **66**: 708–14.

Hay, R. L. & Leakey, M. D. (1982). The fossil footprints of Laetoli. *Scientific American*, **246**(2): 38–45.

Heath, D., Rapp, R. & Taussig, K. (2004). Genetic citizenship. In D. Nugent & J. Vincent, eds. *A Companion to the Anthropology of Politics*. Oxford: Blackwell.

Hebl, M. R., Ruggs, E. N., Singletary, S. L. & Beal, D. J. (2008). Perceptions of obesity across the lifespan. *Obesity*, **16**, Supplement 2: S46–S52.

Henneberg, M. & Lauw, C. (1995). Average menarchal age of higher socio-economic status urban Cape coloured girls assessed by means of status quo and recall methods. *American Journal of Physical Anthropology*, **96**: 1–6.

Herring, D. A., Saunders, S. R. & Katzenberg, M. A. (1998). Investigating the weaning process in past populations. *American Journal of Physical Anthropology*, **105**: 425–39.

Hill, J. O., Catenacci, V. A. & Wyatt, H. R. (2006). Obesity: etiology. In M. E. Shils, M. Shike, A. C. Ross, B. Caballero & R. J. Cousins, eds. *Modern Nutrition in Health and Disease*. 10th edn. Baltimore: Lippincott Williams & Wilkins, pp. 1013–28.

Hill, M. (2002). Skin color and the perception of attractiveness among African Americans: does gender make a difference? *Social Psychology Quarterly*, **65**: 77–91.

Hill, R. (1999). Retina identification. In A. K. Jain, R. Bolle & S. Pankanti, eds. *Biometrics: Personal Identification in Networked Society*. New York: Kluwer Academic Publishers, pp. 123–41.

Himes, C. L. (2000). Obesity, disease, and functional limitations in later life. *Demography*, **37**: 73–82.

Hirschfeld, L. A. (1996). *Race in the Making: Cognition, Culture, and the Child's Construction of Human Kinds*. Cambridge: MIT Press.

Hitsch, G., Hortacsu, A. & Ariely, D. (2004). What makes you click? An empirical analysis of online dating. *UCSC Economics Department Seminars Paper 3*. Available at http://repositories. cdlib. org/ucsc econ seminar/winter2005/3.

Ho, S. Y. W., Phillips, M. J., Cooper, A. & Drummond, A. J. (2005). Time dependency of molecular rate estimations and systematic overestimation of recent divergence times. *Molecular Biology and Evolution*, **22**: 1561–8.

Hochmeister, M. N. (1995). DNA technology in forensic applications. *Molecular Aspects of Medicine*, **16**: 316–437.

Hockey, J. & Draper, J. 2005. Beyond the womb and the tomb: Identity, (Dis) embodiment and the life course. *Body and Society*, **11**: 41–58.

Hogervorst, T., Bouma, H. W. & de Vos, J. (2009). Evolution of the hip and pelvis. *Acta Orthopeadica*, **80**, S336: 1–39.

Hogle, L. F. (2003). Life/time warranty: rechargeable cells and extendable lives. In M. Lock & S. Franklin, eds. *Remaking life and death towards and anthropology of the biosciences*. New Mexico: Santa Fe School of American Research, pp. 61–96.

Hollimon, S. E. (1997). The third gender in native California: two spirit undertakers among the Chumash and their neighbors. In C. Claassen & R. A. Joyce, eds.,

Women in Prehistory: North America and Mesoamerica. Philadelphia: University of Pennsylvania Press, pp. 171–88.

Hollimon, S. E. (2011). Sex and gender in bioarchaeological research. Theory, method and interpretation. In S. C. Agarwal & B. A. Glencross, eds. *Social Bioarchaeology*. Oxford: Wiley-Blackwell, pp. 150–82.

Holowatz, L. A., Thompson-Torgerson, C. S. & Kenney, W. L. (2007). Altered mechanisms of vasodilation in aged human skin. *Exercise and Sports Science Review*, **35**: 119–25.

Holsinger, K. E. (2001). Hardy-Weinberg Law. In S. Brenner & J. H. Miller, eds. *Encyclopedia of Genetics*. New York: Elsevier Science Inc., pp. 912–14.

Hoppa, R. D. & Vaupel, J. W. (2002). *Paleodemography: Age Distributions from Skeletal Samples*. Cambridge University Press.

Hubbell, L. (2009). Creating self and substance in a virtual world. *Surface Design Journal*, **33**: 46–51.

Hummert, M. L. (1994). Physiognomic cues to age and the activation of stereotypes of the elderly in interaction. *International Journal of Aging and Human Development*, **39**: 5–19.

Humphrey, J. H. & Hutchinson, D. L. (2001). Macroscopic characteristics of hacking trauma. *Journal of Forensic Science*, **46**: 228–33

Hutchinson, N. (2006). Disabling beliefs? Impaired embodiment in the religious tradition of the West. *Body and Society*, **12**: 1–23.

Ikehara-Quebral, R. & Douglas, M. T. (1997). Cultural alteration of human teeth in the Mariana Islands. *American Journal of Physical Anthropology*, **104**: 381–91.

Ikegaya, H., Motani, H., Saukko, P. et al. (2007). BK virus genotype distribution offers information of tracing the geographical origins of unidentified cadaver. *Forensic Science International*, **173**: 41–6.

Iyengar, L., Patkunanathan, B., Lynch, O. T. et al. (2006). Aqueous humour and growth factor-induced lens cell proliferation is dependent on MAPK/ERK1/2 and Akt/PI3-K signalling. *Experimental Eye Research*, **83**: 667–78.

Jablonski, N. G. (2004). The evolution of human skin and skin color. *Annual Review of Anthropology*, **33**: 585–623.

Jablonski, N. G. & Chaplin, G. (2000). The evolution of skin colouration. *Journal of Human Evolution*, **39**: 57–106.

Jackson, S. & Scott, S. (2002). Introduction: the gendering of sociology. In S. Jackson & S. Scott, eds. *Gender: A Sociological Reader*. London: Routledge, pp. 1–26.

Jackson, P. & Williams, D. (2006). The intersection of gender, race and SES: health paradoxes. In A. J. Schulz & L. Mullings, eds. *Gender, Race, Class and Health: Intersectional Approaches*. San Francisco: Jossey-Bass, pp. 131–62.

Jaenisch, R., Eggan, K., Humphreys, D., Rideout, W. & Hochedlinger, K. (2002). Nuclear cloning, stem cells, and genomic reprogramming. *Cloning and Stem Cells*, **4**: 389–96.

Jain, A. K., Prabhakar, S. & Pankanti, S. (2002). On the similarity of identical twin fingerprints. *Pattern Recognition*, **35**: 2653–63.

James, C. S. (1939). Footprints and feet of natives of the Solomon Islands. *Lancet*, **2**: 1390–3.

Jantz, R. L. (1992). Modification of the Trotter and Gleser female stature estimation formulae. *Journal of Forensic Sciences*, **37**: 1230–5.

Jantz, R. L., Hunt, D. R. & Meadows, L. (1994). Maximum length of the tibia: how did Trotter measure it? *American Journal of Physical Anthropology*, **93**: 525–8.

Jarman, K. H., Kreuzer-Martin, H. W., Wunschel, D. S. et al. (2008). Bayesian-integrated microbial forensics. *Applied and Environmental Microbiology*, **74**: 3573–82.

Jarvis, G. K. & Northcott, H. C. (1987). Religion and differences in morbidity and mortality. *Social Science and Medicine*, **25**: 813–24.

Jasienska, G. (2009). Low birth weight of contemporary African Americans: an inter-generational effect of slavery? *American Journal of Human Biology*, **21**: 16–24.

Jha, S. and Adelman, M. (2009). Looking for love in all the white places: a study of skin color preference on Indian matrimonial and mate-seeking websites. *Studies in South Asian Film and Media*, **1**: 65–83.

Johnson, R. (2001). The anthropological study of body decoration as art: collective representations and the somatozation of affect. *Fashion Theory*, **5**: 417–34.

Johnston, P. K. & Sabaté, J. (2006). Nutritional implications of vegetarian diets. In M. E. Shils, M. Shike, A. C. Ross, B. Caballero & R. J. Cousins, eds. *Modern Nutrition in Health and Disease*. 10th edn. Philadelphia: Lippincott Williams & Wilkins, pp. 1638–54.

Jones, A. (2007). Identities edge. *Orion Magazine*, www. orionmagazine. org/index. php/articles/article/103/.

Jordanova, L. J. (1989). *Sexual Visions: Images of Gender in Science and Medicine between the Eighteenth and Twentieth Centuries*. University of Wisconsin Press.

Jowett, S. & Ryan, T. (1985). Skin disease and handicap: an analysis of the impact of skin conditions. *Social Science and Medicine*, **20**: 425–9.

Joyce, R. A. (2005). Archaeology of the body. *Annual Review of Anthropology*, **34**: 139–58.

Joyce, R. A. (2008). *Ancient Bodies, Ancient Lives: Sex, Gender, and Archaeology*. London: Thames and Hudson.

Jurmain, R. (1999). *Stories from the Skeleton: Behavioural Reconstruction in Human Osteology*. London: Routledge.

Juzeniene, A., Setlow, R., Porojnicu, A., Stenidal, A. H. & Moan, J. (2009). Development of different human skin colors: a review highlighting photobiological and photobiophysical aspects. *Journal of Photochemistry and Photobiology B, Biology*, **96**: 93–100.

Kahana, T., Hiss, J. & Smith P. (1998). Quantitative assessment of trabecular bone pattern markers for positive identification. *Journal of Forensic Sciences*, **43**: 1144–7.

Kaptoge, S., Jakes, R. W., Dalzell, N. et al. (2007). Effects of physical activity on evolution of proximal femur structure in a younger elderly population. *Bone*, **40**: 506–15.

Karger, B., Rand, S., Fracasso, T. & Pfeiffer, H. (2008). Bloodstain pattern analysis: casework experience. *Forensic Science International*, **181**: 15–20.

Katzenberg, M. A., Herring, A., & Saunders, S. R. (1996). Weaning and infant mortality: evaluating the skeletal evidence. *Yearbook of Physical Anthropology*, **39**: 177–99.

Kaufman, S. R. & Morgan, L. M. (2005). The anthropology of the beginnings and ends of life. *Annual Review of Anthropology*, **34**: 317–41.

Keelaghan, E., Margolis, D., Zhan, M. & Baumgarten, M. (2008). Prevalence of pressure ulcers on hospital admission among nursing home residents transferred to the hospital. *Wound Repair and Regeneration*, **16**: 331–6.

Keita, S. O. Y. & Kittles, R. A. (1997). The persistence of racial thinking and the myth of racial divergence. *American Anthropologist*, **99**: 534–54.

Keith, A. (1925). Report on bones from Saxon graves. *Wiltshire Archaeology*, **43**: 97–101.

Kelley, J. O & Angel, J. L. (1987). Life stresses of slavery. *American Journal of Physical Anthropology*, **74**: 199–221.

Kellinghaus, M., Schulz, R., Vieth, V., et al. (2010). Enhanced possibilities to make statements on the ossification status of the medial clavicular epiphysis using an amplified staging scheme in evaluating thin-slice CT scans. *International Journal of Legal Medicine*, **124**: 231–325.

Kemkes-Grottenthaler, A. (2002). Aging through the ages: historical perspectives on age indicator methods. In R. D. Hoppa & J. W. Vaupel., eds. *Paleodemography*. Cambridge University Press, pp. 48–72.

Kennedy, K. A. R. (1995). But professor, why teach race identification if races don't exist?. *Journal of Forensic Sciences*, **40**: 797–800.

Kennedy, R. B. (1996). Uniqueness of bare feet and its use as a possible means of identification. *Forensic Science International*, **82**: 81–7.

Kennedy, R. B., Chen, S., Pressman, I. S., Yamashita, A. B. & Pressman, A. E. (2005). A large-scale statistical analysis of barefoot impressions. *Journal of Forensic Sciences*, **50**: 1071–80.

Kerr, A. & Shakespeare, T. (2002) *Genetic Politics: From Eugenics to Genome*. New Clarion Press.

Kerridge, I., Lowe, M., Seldon, M., Enno, A. & Deveridge, S. (1997). Clinical and ethical issues in the treatment of a Jehovah's Witness with Acute Myeloblastic Leukemia. *Archives of Internal Medicine*, **157**: 1753–7.

Khrapko, K. & Vijg, J. (2009). Mitochondrial DNA mutations and aging: devils in the details? *Trends in Genetics*, **25**: 91–8.

Kirby, D. A. (2004). Extrapolating race in GATTACA: genetic passing, identity, and the science of race. *Literature and Medicine*, **23**: 184–200.

Klimentidis, Y. C., Miller, G. F. & Shriver, M. D. (2009). Genetic admixture, self-reported ethnicity, self estimated admixture, and skin pigmentation among Hispanics and native Americans. *American Journal of Physical Anthropology*, **138**: 375–83.

Klingerman, K. M. (2006). *Binding Femininity: An Examination of the Effects of Tightlacing on the Pelvis*. Unpublished MA thesis, Louisiana State University.

Knapp, M., Clarke, A. C., Horsburgh, K. A. & Matisoo-Smith, E. A. (2011). Setting the stage: building and working in an ancient DNA laboratory. *Annals of Anatomy*, doi:10. 1016/j. aanat. 2011. 03. 008.

Knudson, K. J. & Stojanowski, C. J. (2008). New directions in bioarchaeology: recent contributions to the study of human social identities. *Journal of Archaeological Research*, **16**: 397–432.

Knudson, K. J. & Stojanowski, C. J. (2009). *The Bioarchaeology of Identity in the Americas*. University of Florida.

Knüsel, C. J. (2002). More Circe than Cassandra: the Princess of Vix in ritualised social context. *European Journal of Archaeology*, **5**: 276–309.

Knuti, K. A., Amrein, P. C., Chabner, B. A., Lynch Jr, T. J. & Penson, R. T. (2002). Faith, identity, and leukemia: when blood products are not an option. *The Oncologist*, **7**: 371–80.

Koch, L. (2009). Eugenics. In P. Atkinson, P. Glasner & M. Lock, eds. *Handbook of Genetics and Society. Mapping the New Genomic Era*. Oxford: Routledge, pp. 437–47.

Komar, D. (2003). Lessons from Srebrenica: the contributions and limitations of physical anthropology in identifying victims of war crimes. *Journal of Forensic Sciences*, **48**: 1–4.

Konigsberg, L. W. & Frankenberg, S. R. (1992). Estimation of age structure in anthropological demography. *American Journal of Physical Anthropology*, **89**: 235–56.

Kooyman, B., Newman, M. E. & Ceri, H. (1992). Verifying the reliability of blood residue analysis on archaeological tools. *Journal of Archaeological Science*, **19**: 265–9.

Krieger, N. (2003). Genders, sexes, and health: what are the connections – and why does it matter? *International Journal of Epidemiology*, **32**: 652–7.

Krieger, N. (2005). Stormy weather: race, gene expression, and the science of health disparities. *American Journal of Public Health*, **95**: 2155–60.

Krieger, N. & Davy Smith, G. (2004). Bodies count and body counts: social epidemiology and embodying inequality. *Epidemiological Review*, **26**: 92–103.

Krieger, N. & Fee, E. (1996). Measuring social inequalities in health in the United States: an historical review 1900-1950. *International Journal of Health Sciences*, **26**: 391–418.

Krishan, K. (2007). Individualizing characteristics of footprints in Gujjars of North India – forensic aspects. *Forensic Science International*, **169**: 137–44.

Krishan, K. (2008). Establishing correlation of footprints with body weight: forensic aspects. *Forensic Science International*, **179**: 63–9.

Kruger, E. Magnet, S., & Van Loon, J. (2008). Biometric revisions of the 'body' in airports and US welfare reform. *Body and Society*, **14**: 99–121.

Kücken, M. & Newell, A. C. (2005). Fingerprint formation. *Journal of Theoretical Biology*, **235**: 71–83.

Kulthanan, T., Techakampuch, S. & Donphongam, N. (2004). A study of footprints in athletes and non-athletic people. *Journal of the Medical Association of Thailand*, **87**: 788–93.

Kunzle, D. (1982). *Fashion and Fetishism: a Social History of the Corset, Tight-Lacing and Other Forms of Body Sculpture in the West*. Rowman and Littlefield.

Kuzawa, C. W. & Sweet, E. (2009). Epigenetics and the embodiment of race: developmental origins of US racial disparities in cardiovascular health. *American Journal of Human Biology*, **21**: 2-15.

Kwate, N. O. A. (2008). Fried chicken and fresh apples: racial segregation as a fundamental cause of fast food density in black neighborhoods. *Health & Place*, **14**: 32–44.

La Fontaine, J. (1986). An anthropological perspective on children in social worlds. In M. Richards & P. Light, eds. *Children of Social Worlds*. Cambridge: Polity Press, pp. 10–30.

Lafrance, M. (2009). Skin and the self: cultural theory and Anglo-American psychoanalysis. *Body and Society*, **15**: 3–24.

Lambert, H. & McDonald, M. (2009). Introduction. In H. Lambert & M. McDonald, eds. *Social Bodies*. New York: Berghahn Books, pp. 1–15.

Landecker, H. (2011). Food as exposure: nutritional epigenetics and the new metabolism. *Biosocieties*, **6**: 167–94.

Lapham, E. V., Kozma, C. & Weiss, J. O. (1996). Genetic discrimination: perspectives of consumers. *Science*, **274**: 621–4.

Laqueur, T. (1990). *Making Sex. Body and Gender from the Greeks to Freud*. Boston: Harvard University Press.

Larsen, C. S. (1995). Biological changes in human populations with the transition to agriculture. *Annual Review of Anthropology*, **24**: 185–213.

Leach, S., Lewis, M., Chenery, C., Müldner, G. & Eckardt, H. (2009). Migration and diversity in Roman Britain: a multidisciplinary approach to the identification of immigrants in Roman York, England. *American Journal of Physical Anthropology*, **140**: 546–61.

Lederman, S. J., Kilgour, A., Kitada, R., Klatzky, R. L. & Hamilton, C. (2007). Haptic face processing. *Canadian Journal of Experimental Psychology*, **61**: 230–41.

Legasse, A. 1973. *Gada*. New York: Free Press.

Le Huray, J. D. & Schutkowski, H. (2005). Diet and social status during the La Tene period in Bohemia: carbon and nitrogen stable isotope analysis of bone collagen from Kutna Hora-Kankar and Radiovesice. *Journal of Anthropological Archaeology*, **24**: 135–47.

Lemelin, P. (2000). Micro-anatomy of the volar skin and interordinal relationships of primates. *Journal of Human Evolution*, **38**: 257–67.

Lemke, T. (2002). Genetic testing, eugenics and risk. *Critical Public Health*, **12**: 283–90.

Levine, R. A. (1998). Child psychology and anthropology: an environmental view. In C. Panter-Brick, ed. *Biosocial Perspectives on Children*. Cambridge University Press, pp. 102–30.

Lewis, J. E., DeGusta, D, Meyer, M. R. et al., (2011). *The Mismeasure of Science: Stephen Jay Gould versus Samuel George Morton on Skulls and Bias*. PLoS Biol 9: e1001071. doi:10. 1371/journal. pbio. 1001071.

Lewis, M. (2007). *The Bioarchaeology of Children*. Cambridge University Press.

Lewis, M. & Rutty, G. (2003). The endangered child: the personal identification of children in forensic anthropology. *Science & Justice*, **43**: 201–9.

Lewontin, R. C. (1972). The apportionment of human diversity. *Evolutionary Biology*, **6**: 381–98.

Lewontin, R. C. (1991). *Biology as Ideology: the Doctrine of DNA*. New York: Harper Perennial.

Lewontin, R. C. (2000). *The Triple Helix: Gene, Organism, and Environment*. Cambridge: Harvard University Press.

Lieberman, L. (1975). The debate over race: a study in the sociology of Knowledge. In A. Montagu, ed. *Race and IQ*. New York: Oxford University Press, pp. 19–41.

Lieberman, L. & Reynolds, L. T. (1978). The debate over race revisited: an empirical investigation. *Phylon*, **39**: 333–43.

Lindemann, M. (1999). Height and our perception of others: an evolutionary perspective. In U. Eiholzer, F. Haverkamp & L. Voss, eds. *Growth, Stature, and Psychosocial Well-Being*. Seattle: Hogrefe & Huber, pp. 121–30.

Linder, M. & Saltzman, C. L. (1998). A history of medical scientists on high heels. *International Journal of Health Services*, **28**: 201–25.

Lippman, A. (1992). Led (astray) by genetic maps: the cartography of the human genome and health care. *Social Science and Medicine*, **35**: 1469–72.

Livingstone, F. B. (1962). On the non-existence of human races. *Current Anthropology*, **3**: 279–81.

Lock, M. (2002). *Twice dead: organ transplants and the reinvention of death*. University of California Press.

Lock, M. (2004). Living cadavers and the calculation of death. *Body and Society*, **10**: 135–52.

Lock, M., Freeman, J. Chilibeck, G., Beveridge, B. & Padolsky, M. (2007). Susceptibility genes and the question of embodied identity. *Medical Anthropology Quarterly*, **21**: 256–76.

Lombroso, C. & Ferrero, G. (1895). *The Female Offender*. Wm. S. Hein.

Long, J. C., Li, J. & Healy, M. E. (2009). Human DNA sequences: more variation and less race. *American Journal of Physical Anthropology*, **139**: 23–34.

Lopes, S. (2010). Test security: defeating the cheats. *Biometric Security Today*, **4**: 9–11.

López, I. R. (2008). "But you don't look Puerto Rican": The buffering effects of ethnic identity on the relation between skin color and self-esteem among Puerto Rican women. *Cultural Diversity & Ethnic Minority Psychology*, **14**, (2), 102–108.

López Jornet, P., Vicente Ortega, V., Yáñez Gascón, J. et al., (2004). Clinicopathological characteristics of tongue piercing: an experimental study. *Journal of Oral Pathology and Medicine*, **33**: 340–5.

Loth, S. R. & Henneberg, M. (2001). Sexually dimorphic mandibular morphology in the first few years of life. *American Journal of Physical Anthropology*, **115**: 179–86.

Loukas, M., Tubbs, R. S., Louis Jr., R. G. et al. (2007). The cardiovascular system in the pre-Hippocratic era. *International Journal of Cardiology*, **120**: 145–9.

Lozada, M. C. (2011). Marking ethnicity through premortem cranial modification among the pre-Inca Chiribaya, Peru. In M. Bonogofsky, ed. *The Bioarchaeology of the Human Head*. University of Florida Press, pp. 228–40.

Lynch, M. & McNally, R. (2009). Forensic DNA databases and biolegality: the co-production of law, surveillance technology and suspect bodies. In P. Atkinson, P. Glasner & M. Lock, eds. *Handbook of Genetics and Society. Mapping the New Genomic Era*. Oxford: Routledge, pp. 283–301.

Mackie, G. (1996). Ending footbinding and infibulation: a convention account. *American Sociological Review*, **61**: 999–1017.

Magin, P. Adams, J., Heading, G., Pond, D. & Smith, W. (2008). Experiences of skin related teasing and bullying and their psychological sequelae: results of a qualitative study. *Scandinavian Journal of Caring Sciences*, **22**: 430–6.

Mann, R. W. (1998). Use of bone trabeculae to establish positive identification. *Forensic Science International*, **98**: 91–9.

Mao, J. (2008). Foot binding: beauty and torture. *The Internet Journal of Biological Anthropology*, **1** (online).

Marie, C. (2008). Presumptive testing and enhancement of blood. In T. Bevel & R. M. Gardner, eds. *Bloodstain Pattern Analysis*. 3rd edn. Boca Raton: CRC Press, pp. 275–96.

Marino, C., Penedo, M. G., Carreira, M. J. & González, F. (2006). Personal authentication using digital retinal images. *Pattern Analysis and Applications*, **9**: 21–33.

Marmot, M. (2006). Introduction. In M. Marmot & R. G. Wilkinson, eds. *Social Determinants of Health*. 2nd edn. Oxford University Press, pp. 1–5.

Marshall, V. W. (1996). The state of theory in aging and the social sciences. In R. H. Binstock & L. K. George, eds. *Handbook of Aging and the Social Sciences*. 4th edn. New York: Academic Press, pp. 12–30.

Martin-McDonald, K., McIntyre, P. & Hegney, D. (2005). Separation of conjoined twins: experiences of perioperative nurses and their recommendations. *International Nursing Review*, **52**: 52–9.

Mathieson, I., Upton, D. & Birchenough, A. (1999). Comparison of footprint parameters calculated from static and dynamic footprints. *The Foot*, **9**: 145–9.

Matsuo, K. & Irie, N. (2008). Osteoclast-osteoblast communication. *Archives of Biochemistry and Biophysics*, **473**: 201–9.

Mays, S. (2005). Paleopathological study of the hallux valgus. *American Journal of Physical Anthropology*, **126**: 139–42.

Mays, S., Brickley, M., & Ives, R. (2006). Skeletal manifestations of rickets in infants and young children in a historic population from England. *American Journal of Physical Anthropology*, **129**: 362–74.

Mays, S. & Cox, M. (2000). Sex determination in skeletal remains. In M. Cox & S. Mays, eds. *Human Osteology in Archaeology and Forensic Science*. London: Greenwich Medical Media Ltd, pp. 117–30.

Mays, S., Ives, R. & Brickley, M. (2009). The effects of socioeconomic status on endochondral and appositional bone growth, and acquisition of cortical bone in children from 19th century Birmingham, England. *American Journal of Physical Anthropology*, **140**: 410–16.

McCabe, L. L. & McCabe, E. R. B. (2008). *DNA: Promise and Peril*. California: University of California Press Ltd.

McCalden, R. W., MacDonald, S. J., Bourne, R. B. & Marr, J. T. (2009). A randomized controlled trial comparing 'High-Flex' *vs* 'Standard' posterior cruciate substituting polyethylene tibial inserts in total knee arthroplasty. *Journal of Arthroplasty*, **24**: 33–8.

McGowan, C. (1999). *A Practical Guide to Vertebrate Mechanics*. Cambridge University Press.

Mead, M. (1973). *Coming of Age in Samoa: A Psychological Study of Primitive Youth for Western Civilisation*. New York: Morrow.

Meier-Augenstein, W. & Fraser, I. (2008). Forensic isotope analysis leads to identification of a mutilated murder victim. *Science & Justice*, **48**: 153–9.

Meijerman, L., Sholl, S., De Conti, F. et al. (2004). Exploratory study on classification and individualisation of earprints. *Forensic Science International*, **140**: 91–9.

Meneghini, R. M., Ford, K. S., McCollough, C. H., Hanssen, A. D. & Lewallen, D. G. (2010). Bone remodeling around porous metal cementless acetabular components. *Journal of Arthroplasty*, **25**: 741–7.

Merbs, C. F. (1983). *Patterns of activity-induced pathology in a Canadian Inuit population*. National Museum of Man Mercury Series 119. Ottawa: Archaeological Survey of Canada.

Meskell, L. (1999). *Archaeologies of Social Life*. Oxford: Blackwell.

Meskell, L. & Preucel, R. W. (2004*). Identities. A Companion to Social Archaeology*. Oxford: Blackwell, pp. 121–41.

Meyers, M. S. & Foran, D. R. (2008). Spatial and temporal influences on bacterial profiling of forensic soil samples. *Journal of Forensic Sciences*, **53**: 652–60.

Millner, V. S., Eichold II, B. H., Sharpe, T. H. & Lynn Jr, S. C. (2005). First glimpse of the functional benefits of clitoral hood piercings. *American Journal of Obstetrics and Gynecology*, **193**: 675–6.

Milner, G. R. & Larsen, C. S. (1991). Teeth as artifacts of human behaviour: intentional mutilation and accidental modification. In M. A. Kelley & C. S. Larsen, eds. *Advances in Dental Anthropology*. New York: Wiley-Liss, pp. 357–78.

Mitchell, P. D., Stern, E., Tepper, Y. (2008). Dysentery in the crusader kingdom of Jerusalem: an ELISA analysis of two medieval latrines in the city of Acre (Israel). *Journal of Archaeological Science*. **35**(7): 1849–53.

Moen, P. (1996). Gender, age and the life course. In R. H. Binstock & L. K. George, eds. *Handbook of Aging and the Social Sciences*. 4th edn. New York: Academic Press, pp. 181–7.

Molleson, T. I. & Cox, M. (1993). *The Spitalfields Project Volume 2: The Anthropology; the Middling Sort*. York: Council for British Archaeology Monograph 86.

Montagu, A. (1978). *Touching: the Human Significance of the Skin*. 2nd edn. London: Harper & Row.

Montepare, J. M. (2006). Body consciousness across the adult years: variations with actual and subjective age. *Journal of Adult Development*, **13**: 102–7.

Montgomery, J. & Evans J. A. (2006). Immigrants on the Isle of Lewis – combining traditional funerary and modern isotope evidence to investigate social differentiation, migration and dietary change in the Outer Hebrides of Scotland. In R. L. Gowland & C. J. Knüsel, eds. *The Social Archaeology of Funerary Remains*. Oxford: Oxbow, pp. 122–42.

Montgomery, J., Evans, J. A., Powlesland, D. & Roberts, C. A. (2005). Continuity or colonisation in Anglo-Saxon England? Isotope evidence for mobility, subsistence practice, and status at West Heslerton. *American Journal of Physical Anthropology*, **126**(2): 123–13.

Moody, J. (2004). Public perceptions of biometric devices: the effect of misinformation on acceptance and use. *Journal of Issues Informing Science and Information*, **1**: 753–61.

Moore, H. L. (1994). *A Passion for Difference: Essays in Anthropology and Gender.* Cambridge: Polity Press.

Morgan, D. (2002). You too can have a body like mine. In S. Jackson & S. Scott, eds. *Gender – A Sociological Reader.* London: Routledge, pp. 406–22.

Morrison, I., Löken, L. S. & Olausson, H. (2010). The skin as a social organ. *Experimental Brain Research,* **204**: 305–14.

Moser, C., Lee, J. & Christensen, P. (1993). Nipple piercing: an exploratory-descriptive study. *Journal of Psychology & Human Sexuality,* **6**: 51–61.

Mpontshane, N., Van den Broeck, J., Chhagan et al. (2008). HIV infection is associated with decreased dietary diversity in South African children. *Journal of Nutrition,* **138**: 1705–11.

Mukhopadhyay, C. C. & Moses, Y. T. (1997). Restablishing race in anthropological discourse. *American Anthropologist,* **99**: 516–33.

Mullings, L. & Schulz, A.J. (2006). Intersectionality and health. In A. J. Schulz & L Mullings, eds. *Gender, Race, Class and Health: Intersectional Approaches.* Jossey-Bass, pp. 3–20.

Muramoto, O. (2001). Bioethical aspects of the recent changes in the policy of refusal of blood by Jehovah's Witnesses. *British Medical Journal,* **322**: 37–9.

Murray, C. D. (2001). The experience of body boundaries by Siamese twins. *New Ideas in Psychology,* **19**: 117–30.

Murray, S. (2009). Marked as pathological: fat bodies as virtual confessors. In J. Wright & V. Harwood, eds. *Biopolitics and the Obesity Epidemic.* New York: Routledge, pp. 78–91.

Myerhoff, B. (1984). Rites and signs of ripening: the intertwining of ritual, time, and growing older. In D. I. Kertzer & J. Keith, eds. *Age and Anthropological Theory.* Ithaca: Cornell University Press, pp. 305–30.

Myers, J. (1992). Nonmainstream body modification: genital piercing, branding, burning and cutting. *Journal of Contemporary Ethnography,* **21**: 267–306.

Nakanishi, H., Kido, A., Ohmori, T., Takada, A. et al. (2009). A novel method for the identification of saliva by detecting oral streptococci using PCR. *Forensic Science International,* **183**: 20–3.

Nazroo, J. Y. & Williams, D. R. (2006). The social determination of ethnic/racial inequalities in health. In M. Marmot & R. G. Wilkinson, eds. *Social Determinants of Health.* 2nd edn. Oxford University Press, pp. 238–66.

Nelkin, D. & Lindee, M. S. (1995). *The DNA Mystique: the Gene as a Cultural Icon.* New York: W. H. Freeman and Company.

Neves, C., Prieto, D., Sola, S. & Antunes, M. J. (2009). Heart transplantation from donors of different ABO blood types. *Transplantation Proceedings,* **41**: 938–40.

Newton, D. E. (2007). *Forensic Chemistry.* New York: Facts on File, Inc.

Novas, C. & Rose, N. (2000). Genetic risk and the birth of the somatic individual. *Economy and Society,* **29**: 485–513.

O'Donnell, C., Iino, M., Mansharan, K., Leditscke, J. & Woodford, N. (2011). Contribution of postmortem multidetector CT scanning to identification of the deceased in a mass disaster: experience gained from the 2009 Victorian bushfires. *Forensic Science International,* **205**: 15–28.

Oakley, A. (1972). *Sex, Gender and Society*. London: Temple Smith.

Ohnuki-Tierney, E. (1994). Brain death and organ transplantation. *Current Anthropology* **35**: 233–42.

Ohto, H., Yonemura, Y., Takeda, J. et al. (2009). Guidelines for managing conscientious objection to blood transfusion. *Transfusion Medicine Reviews*, **23**: 221–8.

Olave, E., Del Sol, M., Gabrielli, C., Mondiola, E. & Rodrigues, C. F. S. (2001). Biometric study of the relationships between palmar neurovascular structures, the flexor retinaculum and the distal wrist crease. *Journal of Anatomy*, **198**: 737–41.

Olze, A., Solheim, T., Schulz, R., Kupfer, M. & Schmeling, A. (2010). Evaluation of the radiographic visibility of the root pulp in the lower third molars for the purpose of forensic age estimation in living individuals. *International Journal of Legal Medicine*, **124**: 183–6.

O'Reilly, W. (2007). The 'Adam' case, London. In T. J. U. Thompson & S. M. Black, eds. *Forensic Human Identification: an Introduction*. Boca Raton: CRC Press, pp. 473–84.

Oriá, R. B., Santana, E. N., Fernandes, M. R., Ferreira, F. V. A. & Brito, G. A. C. (2003). Estudo das alterações relacionadas com a idade ne pele humana, utilizando métodos de histo-morfometria e autofluorescência. *Anais Brasileros de Dermatologia*, **78**: 425–34.

Ormstad, K., Karlsson, T., Enkler, L., Law, B. & Rajs, J. (1986). Patterns in sharp force fatalities: a comprehensive forensic medical study. *Journal of Forensic Science*, **31**: 529–42.

Ortner, D. J. (2003). *Identification of Pathological Conditions in Human Skeletal Remains*. Academic Press.

Orton, C. (1981). *Learning to Live with Skin Disorders*. London: Souvenir Press (Educational and Academic) Ltd.

Östör, Á. (1984). Chronology, category, and ritual. In D. I. Kertzer & J. Keith, eds. *Age and Anthropological Theory*. Ithaca: Cornell University Press.

Oudshoorn, N. (1994). *Beyond the Natural Body: An Archaeology of Sex Hormones*. London: Routledge.

Ousley, S., Jantz, R. & Freid, D. (2009). Understanding race and human variation: why forensic anthropologists are good at identifying race. *American Journal of Physical Anthropology*, **139**: 68–76.

PAHO (2000). *Obesity and poverty – a new public health challenge*. Pan American Health Organization Scientific Publication no. 576. Washington.

Park, J. H., Schaller, M. & Crandall, C. S. (2007). Pathogen-avoidance mechanisms and the stigmatization of obese people. *Evolution and Human Behaviour*, **28**: 410–14.

Park, K. & Nye, R.A. (1991). Destiny is anatomy, review of Laqueur's Making Sex: Body and Gender from the Greeks to Freud. *The New Republic*, **18S**: 53–7.

Parkinson, R. A., Dias, K-R., Horswell, J. et al. (2009). Microbial community analysis of human decomposition in soil. In K. Ritz, L. Dawson & D. Miller, eds. *Criminal and Environmental Soil Forensics*. Springer Science+Business Media B.V., pp. 379–94.

Passalacqua, N. V. (2010). The utility of the Samworth and Gowland age-at-death 'Look-up' tables in forensic anthropology. *Journal of Forensic Sciences*, **55**: 482–7.

Peers, L. (2009). On the treatment of dead enemies: indigenous human remains in Britain in the early twenty-first century. In H. Lambert & M. McDonald, eds. *Social Bodies*. New York & Oxford: Berghahn Press, pp. 77–99.

Pelster, B. (2003). Developmental plasticity in the cardiovascular system of fish, with special reference to the zebrafish. *Comparative Biochemistry and Physiology Part A*, **133**: 547–53.

Pennisi, E. (2009). Neanderthal genomics. Tales of a prehistoric human genome. *Science* **323**: 866–71.

Perry, E. M. (2004). *Bioarchaeology of Labor and Gender in the Prehispanic American Southwest*. PhD Dissertation, University of Arizon.

Peterson, A. (1998). Sexing the body: representations of sex in Gray's anatomy, 1858 to the present. *Body and Society*, **4**: 1–15.

Petjua, M., Suteerayongprasert, A., Thongpudc, R. & Hassirid, K. (2007). Importance of dental records for victim identification following the Indian Ocean tsunami disaster in Thailand. *Public Health*, **121**: 251–7.

Petrofsky, J. S., Prowse, M. & Lohman, E. (2008). The influence of aging and diabetes on skin and subcutaneous fat thickness in different regions of the body. *Journal of Applied Research*, **8**: 55–61.

Picardi, A. & Pasquini, P. (2007). Toward a biopsychosocial approach to skin diseases. In P. Porcelli & N. Sonimo, eds. *Psychological Factors Affecting Medical Conditions*. Basel: Karger, pp. 109–26.

Pinz, A., Bernögger, S., Datlinger, P. & Kruger, A. (1998). Mapping the human retina. *IEEE. Transactions on Medical Imaging*, **17**: 606–19.

Pitts, V. L. (2002). *In the Flesh: the Cultural Politics of Body Modification*. New York: Palgrave.

Pivonka, P., Zimak, J., Smith, D. W. et al. (2008). Model structure and control of bone remodelling: a theoretical study. *Bone*, **43**: 249–63.

Plante, G. E. (2003). Impact of aging on the body's vascular system. *Metabolism*, **52**: 31–5.

Pollock, N. K., Laing, E. M., Baile, C. A. et al. (2007). Is adiposity advantageous for bone strength? A peripheral quantitative computed tomography study in late adolescent females. *American Journal of Clinical Nutrition*, **86**: 1530–8.

Polzer, J., Mercer, S. L. & Goel, V. (2002). Blood is thicker than water: genetic testing as citizenship through familial obligation and the management of risk. *Critical Public Health* **12**: 153–68.

Pomeroy, E., Stock, J., Zakrzewski, S. & Lahr, M. M. (2010). A metric study of three types of artificial cranial deformation from north central Peru. *International Journal of Osteoarchaeology*, **20**: 317–34.

Poster, M. (2004). Desiring information and machines. In R. Mitchell & P. Thurtle, eds. *Data Made Flesh: Embodying information*. London: Routledge, pp. 87–102.

Poynter, F. N. L. (1957). Current thought on Harvey. *The British Medical Journal*, **5030**: 1297–9.

Prestigiacomo, C. J. (2010). Deformations and malformations: the history of induced and congenital skull deformity. *Neurosurgical Focus*, **29** (online).

Prevedorou, E., Díaz-Zorita Bonilla, M., Romero, A. et al. (2010). Residential mobility and dental decoration in early medieval Spain: results from the eighth century site of Plaza del Castilla, Pamplona. *Dental Anthropology*, **23**: 42–52.

Privat, K. L., O'Connell, T. C., & Richards, M. P. (2002). Stable isotope analysis of human and faunal remains from the Anglo-Saxon cemetery at Berinsfield, Oxfordshire: dietary and social implications. *Journal of Archaeological Science*, **29**: 779–90.

Prosser, J. (2001). Skin memories. In S. Ahmed & J. Stacey, eds. *Thinking Through the Skin*. New York: Routledge, pp. 52–68.

Prowse, T. L., Saunders, S. R., Schwarcz, H. P. et al. (2008). Isotopic and dental evidence for infant and young child feeding practices in an Imperial Roman skeletal sample. *American Journal of Physical Anthropology*, **137**: 294–308.

Prowse, T. L., Schwarcz, H. P., Garnsey, P. et al. (2007). Isotopic evidence for age-related immigration to Imperial Rome. *American Journal of Physical Anthropology*, **132**: 510–19.

Prowse, T. L., Schwarcz, H. P., Saunders, S. R., Bondioli, L. & Macchiarelli, R. (2005). Isotopic evidence for age-related variation in diet from Isola Sacra, Italy. *American Journal of Physical Anthropology*, **128**: 2–13.

Pugsley, M. K. & Tabrizchi, R. (2000). The vascular system: an overview of structure and function. *Journal of Pharmacological and Toxicological Methods*, **44**: 333–40.

Quint, E. & Breech, L. (2005). Tattoos: beautification or something else? *Journal of Pediatric and Adolescent Gynecology*, **18**: 129–31.

Race, Ethnicity and Genetics Working Group. (2005). The use of racial, ethnic and ancestral categories in human genetic research. *American Journal of Human Genetics*, **77**: 519–32.

Radoinova, D., Tenekedjiev, K. & Yordanov, Y. (2002). Stature estimation from long bone lengths in Bulgarians. *Homo*, **52**: 221–32.

Rahimi, M., Heng, N. C. K., Kieser, J. A. & Tompkins, G. R. (2005). Genotypic comparison of bacteria recovered from human bite marks and teeth using arbitrarily primed PCR. *Journal of Applied Microbiology*, **99**: 1265–70.

Ramsthaler, F., Kreutz, K. & Verhoff, M.A. (2007). Accuracy of metric sex analysis of skeletal remains using FORDISC based on a recent skull collection. *International Journal of Legal Medicine*, **21**: 447–82.

Rasmussen, S. J. (1987). Interpreting androgynous women: female aging and personhood among the Kel Ewey Tuareg. *Ethnology*, **26**: 17–30.

Rasmussen, S. J. (2000). From childbearers to culture bearers: transition to postchildbearing among Taureg women. *Medical Anthropology*, **19**: 91–116.

Rassmussen, M., Li, Y., Lindgreen, S. et al. (2010). Ancient genome sequence of an extinct palaeo-eskimo. *Nature* **463**: 757–62.

Rathbun, T. A. (1987). Health and disease at a South Carolina plantation. *American Journal of Physical Anthropology*, **74**: 239–53.

Ravelli, A. C. J., van der Meulen, J. H. P., Osmond, C., Barker, D. J. P. & Bleker, O. P. (1999). Obesity at the age of 50 y in men and women exposed to famine prenatally. *American Journal of Clinical Nutrition*, **70**: 811–16.

Rawcliffe, C. (2006). *Leprosy in Medieval England*. London: Boydell Press.

Raxter, M. H., Auerbach, B. M. & Ruff, C. B. (2006). Revision of the Fully technique for estimating statures. *American Journal of Physical Anthropology*, **130**: 374–84.

Rebay-Salisbury, K., Sørensen, M. L. & Hughes, J. (2010). *Body Parts and Bodies Whole*. Oxford: Oxbow.

Reitman, R. D., Emerson, R., Higgins, L. & Head, W. (2003). Thirteen year results of total hip arthroplasty using a tapered titanium femoral component inserted without cement in patients with Type C bone. *Journal of Arthroplasty*, **18**, Supplement 1: 116–21.

Relethford, J. H. (2009). Race and global patterns of phenotypic variation. *American Journal of Physical Anthropology*, **139**: 16–22.

Reynolds, H. M., Dunbar, P. R., Uren, R. F. et al. (2007). Three-dimensional visualisation of lymphatic drainage patterns in patients with cutaneous melanoma. *Lancet Oncology*, **8**: 806–12.

Richards, M. P., Molleson, T. I., Vogel, J. C. & Hedges, R. E. M. (1998). Stable isotope analysis reveals variations in human diet at the Poundbury Camp cemetery site. *Journal of Archaeological Science*, **25**: 1247–52.

Riedler, J., Braun-Fahrländer, C., Eder, W. et al. (2001). Exposure to farming in early life and development of asthma and allergy: a cross-sectional survey. *Lancet*, **358**: 1129–33.

Rifkin, B. A., Ackerman, M. J. & Folkenberg, J. (2006). *Human Anatomy: Depicting the Body from the Renaissance to Today*. London: Thames & Hudson.

Robb, J. (2002). Time and Biography: Osteobiography of the Italian Neolithic lifespan. In Y. Hamilaki, M. Pluciennik, & S. Tarlow, eds. *Thinking Through the Body: Archaeologies of Corporeality*. New York: Kluwer Academic/Plenum, pp. 153–72.

Robb, J. (2009). Towards a critical otziography: inventing prehistoric bodies. In H. Lambert & M. McDonald, eds. *Social Bodies*. New York: Berghahn Books, pp. 100–28.

Robb, J., Bigazzi, R., Lazzari, L., Scarsini, C. & Sonego, F. (2001). Social 'status' and biological 'status': A comparison of grave goods and skeletal indicators from Pontecagnano. *American Journal of Physical Anthropology*, **115**: 213–28.

Robbins, L. M. (1985). *Footprints: Collection, Analysis, and Interpretation*. Springfield: Charles C. Thomas.

Roberti, J. W., Storch, E. A. & Bravata, E. A. (2004). Sensation seeking, exposure to psychological stressors, and body modifications in a college population. *Personality and Individual Differences*, **37**: 1167–77.

Roberts, C. A. & Buikstra, J. E. (2003). *The Bioarchaeology of Tuberculosis: A Global View on a Re-emerging Disease*. Gainesville, Fl: University Press of Florida.

Roberts, C. A. & Manchester, K. (2007). *The Archaeology of Disease*. New York: Cornell University Press.

Robertson, A., Brunner, E. & Sheiham, A. (2006). Food is a political issue. In M. Marmot & R. G. Wilkinson, eds. *Social Determinants of Health*. 2nd edn. Oxford University Press, pp. 172–95.

Robins, A. L. (2009). The evolution of light skin colour: role of vitamin D disputed. *American Journal of Physical Anthropology*, **139**: 447–50.

Rogers, T. F. (2004). Safeguadring Tranquility Base: why the Earth's Moon base should become a World Heritage Site. *Space Policy*, **20**: 5–6.

Rose, N. (2000). The biology of culpability: pathological identity and crime control in a biological culture. *Theoretical Criminology*, **4**: 5–43.

Rose, N. (2001). The politics of life itself. *Theory, Culture and Society*, **18**: 1–30.

Rose, N. & Novas, C. (2005). Biological Citizenship. In A. Ong & S. Collier, eds. *Global Assemblages: Technology, Politics and Ethics as Anthropological Problems*. Oxford: Blackwell, pp. 439–63.

Roseboom, T. J., van der Meulen, J. H. P., Ravelli, A. C. J. et al. (2001). Effects of prenatal exposure to the Dutch famine on adult disease in later life: an overview. *Molecular and Cellular Endocrinology*, **185**: 93–8.

Rothstein, M. A. & Joly, Y. (2009). Genetic information and insurance underwriting. Contemporary issues and approaches in the global economy. In P. Atkinson, P. Glasner & M. Lock, eds. *Handbook of Genetics and Society. Mapping the New Genomic Era*. Routledge: Oxford, pp. 127–44.

Roubertoux, P. L. & Carlier, M. (2011). Good use and misuse of 'genetic determinism'. *Journal of Physiology – Paris*, **105**: 190–4.

Ruder, T. D., Kraehenbuehl, M., Gotsmy, W. F. et al. (2011). Radiologic identification of disaster victims: a simple and reliable method using CT of the paranasal sinuses. *European Journal of Radiology*, **81**(2): e132–8.

Samworth, R. & Gowland, R. L. (2007). Estimation of adult skeletal age-at-death: statistical assumptions and applications. *International Journal of Osteoarchaeology*, **17**: 174–88.

Sarich, V. M. & Miele, F. (2004). *Race: The Reality of Human Differences*. Westview Press.

Sauer, N. J. (1992). Forensic anthropology and the concept of race: If races don't exist, why are forensic anthropologists so good at identifying them? *Social Science & Medicine*, **34**: 107–11.

Sauer, N. J. & Wankmiller, J. C. (2009). The assessment of ancestry and the concept of race. In S. Blau & D. H. Ubelaker, eds. *Handbook of Forensic Anthropology and Archaeology*. California: Left Coast Press, Inc., pp. 187–200.

Saunders, S. R. (2000). Subadult skeletons and growth related studies. In M. A. Katzenberg & S. R. Saunders, eds. *Skeletal Biology of Past Peoples: Research Methods*. New York: Wiley-Liss, pp. 135–61.

Saville, P. A., Hainsworth, S. V. & Rutty, G. N. (2007). Cutting crime: the analysis of the 'uniqueness' of saw marks on bone. *International Journal of Legal Medicine*, **121**: 349–58.

Sawday, J. (1995). *The Body Emblazoned: Dissection and the Human Body in Renaissance Culture*. London: Routledge.

Schaefer, M. C. (2008). A summary of epiphyseal union timings in Bosnian males. *International Journal of Osteoarchaeology*, **18**: 536–45.

Schaumann, B. & Alter, M. (1976). *Dermatoglyphics in Medical Disorders*. New York: Springer-Verlag.

Scheper-Hughes, N. (2000). The global traffic in human organs. *Current Anthropology*, **41**: 191–224.

Scheper-Hughes, N. (2011). Mr Tati's holiday and João's safari – seeing the world through transplant tourism. *Body and Society*, **17**: 55–92.

Scheper-Hughes, N. & Lock, M. M. (1987). The mindful body: a prolegomenon to future work in medical anthropology. *Medical Anthropology Quarterly*, **1**(1): 6–41.

Scheuer, L. & Black, S. (2000). *Developmental Juvenile Osteology*. London: Academic Press.

Schiebinger, L. (1986). Skeletons in the closet: the first illustrations of the female skeleton in eighteenth century anatomy. *Representations*, **14**: 42–83.

Schiebinger, L. (2004). *Nature's Body: Gender in the Making Modern Science*. New Brunswick: Rutgers University Press.

Schildkrout, E. (1978). Age and gender in Hausa society: socioeconomic roles of children in urban Kano. In L. La Fontaine, ed. *Sex and Age as Principles of Social Differentiation*. London: Academic Press, pp. 109–38.

Schildkrout, E. (2004). Inscribing the body. *Annual Review of Anthropology*, **33**: 319–44.

Schillaci, M. & Stojanowski, C. (2002). A reassessment of matrilocality in Chacoan culture. *American Antiquity*, **67**: 343–56.

Schmeling, A., Reisinger, W., Geserick, G. & Olze, A. (2006). Age estimation of unaccompanied minors: Part 1 general considerations. *Forensic Science International*, **159**, Supplement 1: S61–S64.

Schmeling, A., Reisinger, W., Loreck, D. et al. (2000). Effects of ethnicity of skeletal maturation: consequences for forensic age estimates. *International Journal of Legal Medicine*, **113**: 253–8.

Schmidt, S., Kock, B., Schulz, R., Reisinger, W. & Schmeling, A. (2008). Studies in use of the Greulich-Pyle skeletal age method to assess criminal liability. *Legal Medicine*, **10**: 190–5.

Schmitt, A., Murail, P., Cunha, E. & Rougé, D. (2002). Variability of the pattern of aging on the human skeleton: Evidence from bone indicators and implications on age at death estimation. *Journal of Forensic Sciences*, **47**: 1203–9.

Schmidt-Schultz, T. H. & Schultz, M. (2004). Bone protects proteins over thousands of years: extraction, analysis, and interpretation of extracellular matric proteins in archaeological skeletal remains. *American Journal of Physical Anthropology*, **123**: 30–9.

Schneiner, B. (1999). The uses and abuses of biometrics. *Communications of the ACM*: **48**: 136.

Schoeninger, M. J., DeNiro, M. J. & Tauber, H. (1983). Stable isotope ratios of bone collagen reflect marine and terrestrial components of prehistoric human diet. *Science*, **220**: 1381–3.

Schutkowski, H. (1993). Sex determination of infant and juvenile skeletons: I Morphognostic features. *American Journal of Physical Anthropology*, **90**: 199–206.

Schwarcz, H. P. & Schoeninger, M. (1991). Stable isotope analysis in human nutritional ecology. *Yearbook of Physical Anthropology*, **34**: 282–322.

Schweich, M. & Knüsel, C. (2003). Biocultural effects in medieval populations. *Economics and Human Biology*, **1**: 367–77.

Scully, J. L. (2009). Towards a bioethics of disability and impairment. In P. Atkinson, P. Glasner & M. Lock., eds. *Handbook of Genetics and Society. Mapping the New Genomic Era*. Oxford: Routledge, pp. 367–81.

Seiden, S. C. & Morin, K. (2002). The physician as gatekeeper to the use of genetic information in the criminal justice system. *The Journal of Law, Medicine & Ethics*, **30**: 88–94.

Seiter, J. S. & Hatch, S. (2005). Effect of tattoos on perceptions of credibility and attractiveness. *Psychological Reports*, **96**: 1113–20.

Sforza, C., Grandi, G., Catti, F. et al. (2009). Age- and sex-related changes in the soft tissues of the orbital region. *Forensic Science International*, **185**: 1–3.

Shao, J. & Scoggin, M. (2009). Solidarity and distinction in blood: contamination, morality and variability. *Body and Society*, **15**: 29–49.

Shakespeare, T. & Watson, N. (2002). The social model of disability: an outdated ideology? *Research in Social Science and Disability*, **2**: 9–29.

Shapiro, B. & Hofreita, M. (2010). Analysis of ancient human genomes: using next generation sequencing, 20-field coverage of the genome of a 4,000-year-old human from Greenland has been obtained. *Bioessays*, **32**: 388–91.

Sharapova, S. & Razhev, D. (2011). Skull deformation during the Iron Age in the Trans-Urals and Western Siberia. In M. Bonogofsky, ed. *The Bioarchaeology of the Human Head*. University of Florida Press, pp. 202–27.

Sharp, L. (2006). *Strange Harvest: Organ Transplants, Denatured Bodies and the Transformed Self*. University of California Press.

Shaw, M., Dorling, D. & Davey-Smith, G. (2006). Poverty, social exclusion, and minorities. In M. Marmot & R. G. Wilkinson, eds. *Social Determinants of Health*, 2nd edn. Oxford University Press, pp. 196–223.

Shaw, R. B., Katzel, E. B., Koltz, P. F. et al. (2010) Aging of the facial skeleton: aesthetic implications and rejuvenation strategies. *Plastic Reconstructive Surgery*, **127**: 374–83.

Sheldon, S. (2002). Masculine Body. In M. Evans & E. Lee, eds. *Real Bodies: A Sociological Introduction*. New York: Palgrave, pp. 14–28.

Shendure, J. & Ji, H. (2008). Next-generation DNA sequencing. *Nature Biotechnology*, **26**: 1135–45.

Shilling, C. (1993). *The Body in Social Theory*. London: Sage.

Shim, J. K. (2005). Constructing race along the science-lay divide: racial formation in the epidemiology and experience of cardiovascular disease. *Social Studies of Science*, **35**: 405–36.

Shuler, K. A. (2011). Life and death on a Barbadian sugar plantation: historic and bioarchaeological views for infection and mortality at Newton plantation. *International Journal of Osteoarchaeology*, **21**: 66–81.

Sicherer, S. H. & Leung, D. Y. M. (2007). Advances in allergic skin disease, anaplylaxis, and hypersensitivity reactions to foods, drugs, and insects. *Journal of Allergy and Clinical Immunology*, **119**: 1462–9.

Simon, C. & Goldstein, I. (1935). A new scientific method of identification. *New York State Journal of Medicine*, **35**: 901–6.

Sinden, R. R. (1994). *DNA Structure and Function*. San Diego: Academic Press, Inc.

Skeggs, B. (1997). *Formations of Class and Gender*. London: Sage.

Skinner, M., Alempijevic, D. & Djuric-Srejic, M. (2003). Guidelines for international forensic bio-archaeology monitors of mass grave exhumations. *Forensic Science International*, **134**: 81–92.

Smay, D. & Armelagos, G. (2000). Galileo wept: a critical assessment of the use of race in forensic anthropology. *Transforming Anthropology*, **9**: 19–29.

Smedley, A. (2007). *Race in North America. Origin and Evolution of a Worldview.* Boulder: Westview Press.

Smith, P. R. & Wilson, M. T. (1992). Blood residues on ancient tool surfaces: a cautionary note. *Journal of Archaeological Science*, **19**: 237–41.

Smith, P. R. & Wilson, M. T. (2001). Blood residues in archaeology. In D. R. Brothwell & A. M. Pollard, eds. *Handbook of Archaeological Sciences*. England: John Wiley and Sons, Ltd., pp. 313–22.

Smith, V. A., Christensen, A. M. & Myers, S. W. (2010). The reliability of visually comparing small frontal sinuses. *Journal of Forensic Sciences*, **55**: 1413–15.

Smoyer-Tomic, K. E., Spence, J. C., Raine, K. D. et al. (2008). The association between neighborhood socioeconomic status and exposure to supermarkets and fast food outlets. *Health & Place*, **14**: 740–54.

Sofaer, J. (2006). *The Body as Material Culture: A Theoretical Osteoarchaeology*. Cambridge University Press.

Sofaer, J. (2011). Towards a social bioarchaeology of age. In S. C. Agarwal & B. A. Glencross, eds. *Social Bioarchaeology*. Oxford: Wiley-Blackwell, pp. 285–311.

Sofaer Derevenski, J. (1997). Age and gender at the site of Tiszapolgár-Basatanya, Hungary. *Antiquity*, **71**: 875–89.

Sofaer Derevenski, J. (2000). Sex differences in activity-related osseous change in the spine and the gendered division of labor at Ensay and Wharram Percy, UK. *American Journal of Physical Anthropology*, **111**: 333–54.

Sørensen, M. L. S. (2000). *Gender Archaeology*. Cambridge: Polity Press.

Spannier, B. B. (1995). *Im/Partial Science: Gender Ideology in Molecular Biology*. Indianapolis: University of Indiana Press.

Spennemann, D. H. R. (2004). The ethics of treading on Neil Armstrong's footprints. *Space Policy*, **20**: 279–90.

Spennemann, D. H. R. (2006). Out of this World: issues of managing tourism and humanity's heritage on the Moon. *International Journal of Heritage Studies*, **12**: 356–71.

Spindler, K. (1994). *The Man in the Ice. The Preserved Body of a Neolithic Man Reveals the Secrets of the Stone Age*. London: Phoenix.

Sprung, J., Kindscher, J. D., Wahr, J. A. et al. (2002). The use of bovine Hemoglobin Glutamer-250 (Hemopure®) in surgical patients: results of a multicenter, randomized, single-blinded trial. *Anesthesia and Analgesia*, **94**: 799–808.

Steadman, D. W. & Haglund, W. D. (2005). The scope of anthropological contributions to human rights investigations. *Journal of Forensic Sciences* **50**: 23–30.

Steckel, R. H. (2009). Heights and human welfare: recent developments and new directions. *Explorations in Economic History*, **46**: 1–23.

Steele, V. (2001). *The Corset: A Cultural History.* New Haven: Yale University Press.

Steinberg, D. L. (2000). 'Recombinant bodies': narrative, metaphor and the gene. In. S. J. Williams, J. Gabe & M. Calnan, eds. *Health, Medicine and Society. Key Theories, Future Agendas.* London, Routledge, pp. 146–68.

Stern, M. (2008). 'Yes: – no: – I have been sleeping – and now – now – I am dead': undeath, the body and medicine. *Studies in History, Philosophy, Biology and Biomedical Science*, **39**: 347–54.

Stewart, S., Hansen, T. S. & Carey, T. A. (2010). Opportunities for people with disabilities in the virtual world of *Second Life. Rehabilitation Nursing*, **35**: 254–9.

Stojanowski, C. & Schillaci, M. A. (2002). A reassessment of matrilocality in Chacan culture. *American Antiquity*, **67**: 343–56.

Stolberg, M. (2003). A woman down to her bones: the anatomy of sexual difference in the 16th and early 17th centuries. *Isis*, **94**: 274–99.

Stone, L., Lurquin, P. F. & Cavalli-Sforza, L. L. (2007). *Genes, Culture, and Human Evolution: A Synthesis.* Oxford: Blackwell Publishing Ltd.

Stone, P. & Walrath, D. (2006). The gendered skeleton: anthropological interpretations of the bony pelvis. In R. L. Gowland & C. J. Knüsel, eds. *The Social Archaeology of Funerary Remains.* Oxford: Oxbow Books, pp. 168–78.

Strathern, M. (2009). Afterword. *Body and Society*, **15**: 217–22.

Stringer, C. (2011) Rethinking 'Out of Africa'. Edge. www.edge.org/conversation. php?cid=rethinking-out-of-Africa. Accessed 25/08/12.

Strong, T. (2009). Vital publics of pure blood. *Body and Society*, **15**: 169–91.

Sullivan, A. (2004). Reconstructing relationships among mortality, status, and gender at the Medieval Gilbertine Priory of St Andrew, Fishergate, York. *American Journal of Physical Anthropology*, **124**: 330–45.

Sung, Tz'u. (1247). *Collected Writings on the Washing Away of Wrongs.* Translated by B. E. McKnight (1981). Centre for Chinese Studies: University of Michigan.

Sutanto, J., Phang, C. W., Tan, C. H. & Lu, X. (2011). Dr. Jekyll vis-à-vis Mr. Hyde: Personality variation between virtual and real worlds. *Information & Management*, **48**: 19–26.

Sweet, E., McDade, T. W., Kiefe, C. I. & Liu, K. (2007). Relationships between skin colour, income, and blood pressure among African Americans in the CARDIA study. *American Journal of Public Health*, **97**: 2253–9.

Szreter, S. (2005). *Health and Wealth.* New York: University of Rochester Press.

Tate, S. (2001). 'That is my star of David': skin, abjection and hybridity. In S. Ahmed & J. Stacey, eds. *Thinking Through the Skin.* New York: Routledge, pp. 209–22.

Tarvis, C. (1992). *The Mismeasure of Woman.* New York: Simon and Schuster.

Thomas, J. (2007). Archaeology's humanism and the materiality of the body. In T. Insoll. ed. *Archaeology of Identities: A Reader.* New York: Routledge, pp. 211–24.

Thompson, T. J. U. (2001). Legal and ethical considerations of forensic anthropological research. *Science & Justice*, **41**: 261–70.

Thompson, T. J. U. & Black, S. M. (2007). *Forensic Human Identification: an Introduction*. Boca Raton: CRC Press.

Thompson, T. J. U. & Inglis, J. (2009). Differentiation of serrated and non-serrated blades from stab marks in bone. *International Journal of Legal Medicine*, **123**: 129–35.

Thompson, T. J. U. & Puxley, A. (2007). Personal effects. In T. J. U. Thompson & S. M. Black, eds. *Forensic Human Identification: An Introduction*. Boca Raton: CRC Press, pp. 365–77.

Thurtle, P. & Mitchell, R. (2004). Data made flesh: the material poiesis of informatics. In R. Mitchell & P. Thurtle, eds. *Data Made Flesh: Embodying Information*. London: Routledge, pp. 1–25.

Tiesler, V. (2012). Studying cranial vault modifications in ancient Mesoamerica. *Journal of Anthropological Sciences*, **90**: 1–26.

Tiggemann, M. & Golder, F. (2006). Tattooing: an expression of uniqueness in the appearance domain. *Body Image*, **3**: 309–15.

Timmermans, S. (2006). *Postmortem: How Medical Examiners Explain Suspicious Deaths*. The University of Chicago Press.

Torres-Rouff, C. & Yablonsky, L. T. (2005). Cranial vault modification as a cultural artefact: a comparison of the Eurasian steppes and the Andes. *Homo – Journal of Comparative Human Biology*, **56**: 1–16.

Tower, P. (1955). The *fundus oculi* in monozygotic twins. *Archives of Ophthalmology*, **54**: 225–39.

Tringham, G. M., Nawaz, T. S., Holding, S., McFarlane, J. & Lindow, S. W. (2011). Introduction of first trimester combined test increases uptake of Down's syndrome screening. *European Journal of Obstetrics & Gynecology and Reproductive Biology*, **159**: 95–8.

Trotter, M. & Gleser, G. (1952). Estimation of stature from long bones of American whites and Negroes. *American Journal of Physical Anthropology*, **10**: 469–514.

Trotter, M. & Gleser, G. (1958). A re-evaluation of estimation of stature based on measurements taken during life and the long bones after death. *American Journal of Physical Anthropology*, **16**: 79–123.

Turner, B. S. (1991). Recent developments in the theory of the body. In M. Featherstone, ed. *Body: Social Processes and Cultural Theory*, London: Sage, pp. 1–35.

Turner, B. S. (1995). Aging and identity. Some reflections on the somatization of the self. In M. Featherstone & A. Wernick, eds. *Images of Ageing: Cultural Representations of Later Life*, London: Routledge, p. 149.

Turner, B. S. (1996). *The Body & Society* (2nd ed). London: Sage Publications.

Turner, T. (1980). The social skin. In J. Cherfas & R. Lewin., eds. *Not Work Alone: A Cross-cultural View of Activities Superfluous to Survival*. California: Sage, pp. 112–40.

Turney, J. & Balmer, B. (2000). The genetic body. In R. Cooter & J. Pickstone., eds. *Medicine in the Twentieth Century*. London: Harwood, pp. 399–416.

Tuttle, R., Webb, D., Weidl, E. & Baksh, M. (1990). Further progress on the Laetoli Trails. *Journal of Archaeological Science*, **17**: 347–62.

Twine, R. (2002). Physiognomy, phrenology, and the temporality of the body. *Body and Society*, **8**: 67–88.

Ubelaker, D. H. & Jacobs, C. H. (1995). Identification of orthopaedic device manufacturer. *Journal of Forensic Sciences*, **40**: 168–70.

Ubelaker, D. H. & Zarenko, K. M. (2011). Adipocere: What is known after over two centuries of research. *Forensic Science International*, **208**: 167–72.

Underwood, G. & Batt, V. (1996). *Reading and Understanding*. Oxford: Blackwell Publishers.

Ulijaszek, S. J., Johnston, F. E. & Preece, M. A. (2000). *The Cambridge Encyclopaedia of Human Growth and Development*. Cambridge University Press.

Ullrich, H. E. (2010). Is beauty skin deep? The impact of 'beautiful attributes' on life opportunities and interpersonal relationships: a tale of two sisters in South India. *Journal of the American Academy of Psychoanalysis and Dynamic Psychiatry*, **38**: 243–53.

Urry, S. R. & Wearing, S. C. (2005). Arch indexes from ink footprints and pressure platforms are different. *The Foot*, **15**: 68–73.

Usher, B. M. (2002). Reference samples: the first step in linking biology and age in the human skeleton. In R. D. Hoppa & J. W. Vaupel, eds. *Paleodemography: Age Distributions from Skeletal Samples*. Cambridge University Press, pp. 29–47.

Vail, D. A. (1999). Tattoos are like potato chips … you can't have just one: the process of becoming and being a collector. *Deviant Behaviour: An Interdisciplinary Journal*, **20**: 253–73.

Valéry, P. (1933). *L'idée Fixe*. Paris: Gallimard.

Van Daal, A. (2008). The genetic basis of human pigmentation. *Forensic Science International: Genetic Supplement Series*, **1**: 541–3.

Van Gelder, L. & Sharpe, K. (2009). Women and girls as upper palaeolithic cave artists: deciphering the sexes of finger fluters in Rouffignac cave. *Oxford Journal of Archaeology* **28**(4): 323–33.

Versalius, A. (1543). *Epitome*. Translated by L. R. Lind (1969). Cambridge: MIT Press.

Victorino, C. C. & Gauthier, A. H. (2009). The social determinants of child health: variations across health outcomes: a population-based cross-sectional analysis. *BMC Pediatrics*, **9**: 53–64.

Vincent, S. J. (2009). *The Anatomy of Fashion. Dressing the Body from the Renaissance to Today*. Oxford: Berg.

Vioux-Chagnoleau, C., Lejeune, F., Sok, J., Pierrard, C., Marionnet, C. & Bernerd, F. (2006). Reconstructed human skin: from photodamage to sunscreen photoprotection and anti-aging molecules. *Journal of Dermatological Science Supplement*, **2**: S1–S12.

Virkler, K. & Lednev, I. K. (2009). Analysis of body fluids for forensic purposes: from laboratory testing to non-destructive rapid confirmatory identification at a crime scene. *Forensic Science International*, **188**: 1–17.

Vlak, D., Roksandic, M. & Schillaci, M. A. (2008). Greater sciatic notch as a sex indicator in juveniles. *American Journal of Physical Anthropology*, **137**: 309–15.

Vogel, J. C. & van der Merwe, N. J. (1977). Isotopic evidence for early Maize cultivation in New York state. *American Antiquity*, **42**: 238–42.

Vogel, P. (2005). The current molecular phylogeny of Eutherian mammals challenges previous interpretations of placental evolution. *Placenta*, **26**: 591–6.

von Luschan, F. (1897). *Beiträge zur Völkerkunde der Deutschen Schutzgebieten*. Berlin: Deutsche Buchgemeinschaft.

Wadden, T. A, Byrne, K. J. & Krauthamer-Ewing, S. (2006). Obesity: management. In M. E. Shils, M. Shike, A. C. Ross, B. Caballero & R. J. Cousins, eds. *Modern Nutrition in Health and Disease*. 10th edn. Baltimore: Lippincott Williams & Wilkins, pp. 1029–42.

Wadman, M. (1998). Jewish leaders meet NIH chiefs on genetic stigmatization fears. *Nature*, **392**: 851.

Wadsworth, M. & Butterworth, S. (2006). Early life. In M. Marmot & R. G. Wilkinson, eds. *Social Determinants of Health*. 2nd edn. Oxford University Press, pp. 31–53.

Walby, K. & Carrier, N. (2010). The rise of biocriminology: capturing observable bodily economies of 'criminal man'. *Criminology & Criminal Justice*, **10**: 261–85.

Walker, P. L. (1995). Problems of preservation and sexism in sexing: some lessons from historical collections for palaeodemographers. In S. R. Saunders & A. Herring, eds. *Grave Reflections: Portraying the Past Through Cemetery Studies*. Toronto: Canadian Scholars Press, pp. 31–47.

Walker, P. L. (2005). Greater sciatic notch morphology: sex, age, and population difference. *American Journal of Physical Anthropology*, **127**: 385–91.

Walker, P. L. (2008). Sexing skulls using discriminant function analysis of visually assessed traits. *American Journal of Physical Anthropology*, **136**: 39–50.

Walker, P. L. & Collins Cook, D. (1998). Gender and sex: Vive la difference. *American Journal of Physical Anthropology*, **106**: 255–9.

Walrath, D. E., Turner, P., & Bruzek, J. (2004). Reliability test of the visual assessment of cranial traits for sex determination. *American Journal of Physical Anthropology*, **125**: 132–7.

Wang, L., Leedham, G & Choa, D. S-Y. (2008). Minutiae feature analysis for infrared hand vein pattern biometrics. *Pattern Recognition*, **41**: 920–9.

Warren, M. W. (2007). Interpreting gunshot wounds in the Balkans: evidence for genocide. In Brickley, M. B. & Ferllini, R., eds. *Forensic Anthropology: Case Studies from Europe*. Springfield: Charles C. Thomas Publisher, Ltd., pp. 151–64.

Weaver, D. S. (1980). Sex differences in the ilia of a known sex and age sample of fetal and infant skeletons. *American Journal of Physical Anthropology*, **52**: 191–6.

Weiss, K. (1972). On the systematic bias in skeletal sexing. *American Journal of Physical Anthropology*, **37**: 239–50.

Weiss, S. J. (2005). Haptic perception and the psychosocial function of preterm, low birth weight infants. *Infant Behaviour and Development*, **28**: 329–59.

Wertz, D. C. (2002). Embryo and stem cell research in the United States: history and politics. *Gene Therapy*, **9**: 674–8.

White, C. D. (2005). Gendered food behaviour among the Maya: time, place, status and ritual. *Journal of Social Archaeology*, **5**: 356–82.

White, C. D. & Schwarcz, H. P. (1989). Ancient Maya diet: as inferred from isotopic and elemental analysis of human bone. *Journal of Archaeological Science*, **16**: 451–74.

White, T. D. & Folkens, P. A. (2005). *The Human Bone Manual*. New York: Elsevier Academic Press.

Whitehead, M. (1997). Life and death over the millennium. In Drever, F. & Whitehead, M., eds., *Health Inequalities: Decennial Supplement. Office for National Statistics Series DS #15*. London: HMSO, pp. 7–28.

Wildes, R. P. (1997). Iris recognition: an emerging biometric technology. *Proceedings of the IEEE*, **85**: 1348–63.

Wilkinson, C. M. (2004). *Forensic Facial Reconstruction*. Cambridge University Press.

Wilkinson, R. G. (1999). Putting the picture together: prosperity, redistribution, health, and welfare. In M. Marmot & R. G. Wilkinson, eds. *Social Determinants of Health*. 1st edn. Oxford University Press, pp. 256–74.

Wilkinson, R. G. (2006). Ourselves and others – for better or worse: social vulnerability and inequality. In M. Marmot & R. G. Wilkinson, eds. *Social Determinants of Health*. 2nd edn. Oxford University Press, pp. 341–57.

Willcox, A. W. (2002). Mummies and molecules: molecular biology meets paleopathology. *Clinical Microbiology Newsletter*, **24**: 57–60.

Williams, F. L., Belcher, R. L. & Armelagos, G. J. (2005). Forensic misclassification of ancient Nubian crania: implications for assumptions about human variation. *Current Anthropology*, **46**: 340–6.

Williams, R. & Johnson, P. (2008). *Genetic Policing: the Use of DNA in Criminal Investigations*. Willan Publishing.

Williams, S. E. & Slice, D. E. (2010). Regional shape change in adult facial bone curvature with age. *American Journal of Physical Anthropology*, **143**: 437–47.

Williams, S. J. & Bendelow, G. (1998). *The Lived Body: Sociological Themes, Embodied Issues*. London: Routledge.

Wilson, L. A. B., Cardoso, H. F. V. & Humphrey, L. T. (2011). On the reliability of a geometric morphometric approach to sex determination: a blind test of six criteria of the juvenile ilium. *Forensic Science International*, **206**: 35–42.

Wilson, P. (1999). *Surgery, Skin and Syphilis. Daniel Turner's London (1667-1741)*. Amsterdam/Atlanta, GA: Rodopi Press.

Witz, A. (2000). Whose body matters? Feminist sociology and the corporeal turn in sociology and feminism. *Body and Society*, **6**: 1–24.

Wonder, A. Y. (2007). *Bloodstain Pattern Evidence: Objective Approaches and Case Applications*. New York: Elsevier Academic Press.

Woodward, K. L. (1993). The relationship between skin compliance, age, gender, and tactile discriminative thresholds in humans. *Somatosensory and Motor Research*, **10**: 63–7.

Wright, R. A. & Miller, J. M. (1998). Taboo until today? The coverage of biological arguments in criminology textbooks, 1961 to 1970 and 1987 to 1996. *Journal of Criminal Justice*, **26**: 1–19.

Yang, D. Y. & Watt, K. (2005). Contamination controls when preparing archaeological remains for ancient DNA analysis. *Journal of Archaeological Science*, **32**: 331–6.

Zana, F. & Klein, J. C. (1999). A multimodal registration algorithm of eye fundus images using vessels detection and hough transform. *IEEE Transactions on Medical Imaging*, **18**: 419–28.

Zhou, Z., Jin, X-L., Vogel, D. R., Fang, Y. & Chen, X. (2011). Individual motivations and demographic differences in social virtual world users: an exploratory investigation in *Second Life*. *International Journal of Information Management*, **31**: 261–71.

Index